Markets, Places, Cities

Using a transnational analytical framework, this book provides a comprehensive overview of formal and informal markets and place in globalised cities. It examines how urban markets are situated within social, cultural and media discourses, and within material and symbolic economies. The book addresses four key narratives – redevelopment and relocation; privatization of public space; urban renewal; and urbanism and sustainability – to investigate shared and individual attributes of markets and place in diverse, international urban contexts. With case studies in Sydney, Hong Kong, Beijing, Rio de Janeiro, London, Antwerp, Amsterdam, Paris and San Francisco, experiences of market, place and city are explored through interdisciplinary and multimodal perspectives of visual culture, spatial practice, urban design and textual analysis.

Kirsten Seale lectures in Interdisciplinary Design at the University of Technology Sydney and is Adjunct Fellow at the Institute for Culture and Society, Western Sydney University.

Routledge Studies in Urbanism and the City

This series offers a forum for original and innovative research that engages with key debates and concepts in the field. Titles within the series range from empirical investigations to theoretical engagements, offering international perspectives and multidisciplinary dialogues across the social sciences and humanities, from urban studies, planning, geography, geohumanities, sociology, politics, the arts, cultural studies, philosophy and literature.

Published:

The Urban Political Economy and Ecology of Automobility
Driving cities, driving inequality, driving politics
Edited by Alan Walks

Cities and Inequalities in a Global and Neoliberal World
Edited by Faranak Miraftab, David Wilson and Ken E. Salo

Beyond the Networked City
Infrastructure Reconfigurations and Urban Change in the North and South
Edited by Olivier Coutard and Jonathan Rutherford

Technologies for Sustainable Urban Design and Bioregionalist Regeneration
Dora Francese

Markets, Places, Cities
Kirsten Seale

Forthcoming:

Mega-Urbanization in the Global South
Fast Cities and New Urban Utopias of the Postcolonial State
Edited by Ayona Datta and Abdul Shaban

Markets, Places, Cities

Kirsten Seale

Routledge
Taylor & Francis Group

LONDON AND NEW YORK

First published 2016 by Routledge

2 Park Square, Milton Park, Abingdon, Oxfordshire OX14 4RN
711 Third Avenue, New York, NY 10017

Routledge is an imprint of the Taylor & Francis Group, an informa business

First issued in paperback 2018

British Library Cataloguing in Publication Data
A catalogue record for this book is available from the British Library

Library of Congress Cataloguing in Publication Data
Names: Seale, Kirsten, author.
Title: Markets, places, cities / Kirsten Seale.
Description: New York : Routledge, 2016. |
Series: Routledge studies in urbanism and the city |
Includes bibliographical references. Identifiers: LCCN 2015043513 |
ISBN 9781138839823 (hardback) | ISBN 9781315733234 (e-book)
Subjects: LCSH: Farmers' markets–Case studies. | Urbanization.
Classification: LCC HD9000.5.S383 2016 | DDC 381/.1–dc23
LC record available at http://lccn.loc.gov/2015043513

ISBN: 978-1-138-83982-3 (hbk)
ISBN: 978-1-138-54643-1 (pbk)

Typeset in Times New Roman
by Out of House Publishing

For Nick

Contents

Figures

Acknowledgements

Making a book and watching your research take shape in the written word is alternatively exhilarating and daunting. It is often a solitary task too, and I would like to thank the following colleagues and friends (some of whom are both) for the conversations about this book, and for also reminding me that there is a world beyond its pages: Katie Bryant, Alexandra Crosby, Stephanie Hemelryk Donald, Aslı Duru, Clifton Evers, Glen Fuller, Melinda Harvey, Paula Llavallol, Francis Maravillas, Catriona Menzies-Pike, Sarah Pink, Emily Potter, Carl Power and Amanda Williams. Faye Leerink at Routledge was a patient and helpful contact during the commissioning and production phases. Some of this book was written in the University of Technology Sydney library, which is housed in a former hall of Sydney's central produce market. The library's collections were a wonderful resource, and the members of staff were very helpful in finding material on markets from elsewhere. Thank you to Freek Janssens and Ceren Sezer for inviting me to come and talk with you about markets in Amsterdam. I am very grateful to my family – David, Paul and Rosemary Seale, Nyree Phillips, and Nicholas and Louis Mariette – for continuingly engaging with and showing interest in this topic, and for their company in visiting markets in cities around the world. Nick, in particular, has been the source of much love and practical support, and it is to him that I dedicate this book.

1 What is a market?

Many people have been to a street, open-air, covered, farmers' or second-hand market, or have at least seen one. Every culture has them (Slater and Tonkiss, 2001). They are accessed by and accessible to every sector of society, which is a singular trait in a segmented global consumer landscape. Markets have a ubiquitous universality in spite of their many particularities. When I posed the question I ask in this chapter – 'What is a market?' – to one of my students, she answered, 'I don't know, but I know what it looks like in my mind'. How, then, do we determine that a collection of informal vendors selling *ceviche* out of shopping trolleys on a pedestrian bridge in Santiago is a market, whereas a converted turn-of-the-century tram shed selling street food in Amsterdam is not?

In trying to pin down what a market is, or is not, it is not helpful that some markets are outdoors and some are indoors. Some have permanent infra-structure, while others are itinerant. They can sell expensive vintage objects, or junk. They can be licensed, or unlicensed, or both. The range of variables and contingencies at work in markets is vast and almost unnavigable. Even within consumer ecologies, the position of markets is uncertain, as Pierre Mayol notes:

> At the same time as it is a place of business, it is a place of festival (in small provincial towns, the 'pompom of music' frequently accompanies the weekly markets), halfway between the small shops on the street and the department store, or the supermarket, without the elements that con-stitute it being reabsorbed in one or the other of these terms. It offers a profusion of consumer goods surpassing what a shopkeeper offers, but without falling into the 'distributionalism' of supermarkets.
>
> (de Certeau et al., 1998, p. 107)

Furthermore, markets as a signifier mean different things depending on our subject placement or orientation. They can engender mixophilia or mixophobia (Bauman, 2011). Some people might enjoy the 'thrownto-getherness' (Massey, 2005, p. 151) that occurs at markets. Others might find their jumble of heterogeneous flows and matter viscerally confront-ing, and even disgusting. For some, an everyday market in an ungentrified neighbourhood is to be avoided because they dislike the sounds, smells, sights and tastes encountered there, whereas others might actively seek out this market as an authentic representation of place in the city. In some cities, where affluent urban consumers are embracing the ethos of localism, ethical consumption and sustainability, farmers' markets are resurgent (Lewis and Potter, 2011). In other cities, farmers' markets are a valuable resource for marginalized communities who would not otherwise have access to affordable and healthy fresh produce. Depending on your perspective, art, craft and maker markets are a mode of consumption that communicates individualism and evades the standardization of mass production, or, they are a showcase for the ubiquity of globalized Creative City discourse (Mould, 2015). Urban flea markets are a source of livelihood and inexpensive goods for those on the social and spatial periphery, or, they are instruments for the accumulation of cultural capital, thanks to the fetishization of vintage aesthetics.

The transdisciplinary and geographic breadth of existing research on markets also indicates their complexity as an object of inquiry. A sample of the literature reveals research coming out of sociology (Watson, 2006, 2009; Watson and Studdert, 2006; Watson and Wells, 2005); anthropology (Black, 2012; Lyon and Back, 2012); gender studies (Clark, 2010); urban studies (Dines and Cattell, 2006; Janssens and Sezer, 2013); cultural geography (Bubinas, 2011; Coles and Crang, 2011; Law, 2011); urban design and architecture (Franck, 2005; Parham, 2012, 2015); and marketing (Visconti et al., 2014).

Markets and the market

One aspect that has been largely overlooked in the literature, and a challenge I have taken up in this book, is the need for a conceptual framework to distinguish the types of markets that exist in cities from the overweening concept of 'the market' that is forefront in neo-liberal capitalist political economies. 'Market' as a word and as a concept is polyvalent. Nonetheless, Ellen Meiksins Wood in her study of the origin of capitalism declares market 'the word that lies at the very heart of capitalism' (2002, p. 6). Patrik Aspers (2011) states that the market is forefront and central in capitalist political, economic and social systems. Wood and Aspers are talking about what Slater and Tonkiss (2001) call 'market society', where the logic of the capitalist market has superseded all other economies, dictating and co-opting many individual and collective social transactions. Slater and Tonkiss are noteworthy in drawing attention to the difference between markets and market society: ' "Market-places" are visible public events that happen at a regular time and place, with buildings, rules, governing institutions and other social structures… The spatial and temporal location of marketplaces is a crucial feature' (2001, p. 9). Wood, too, draws a distinction between a market that might be run as a capitalist enterprise and capitalist society where everything is run according to market logic:

Markets of various kinds have existed throughout recorded history and no doubt before, as people have exchanged and sold their surpluses in many different ways and for many different purposes. But the market in capitalism has a distinctive, unprecedented function. Virtually everything in capitalist society is a commodity produced for the market... This market dependence gives the market an unprecedented role in capitalist societies, as not only a simple mechanism of exchange or distribution but the principal determinant and regulator of social reproduction.

(Wood, 2002, pp. 96–97)

More often than not in discussions of markets, abstract and actualized forms tend to get lumped together (Aspers, 2011; Calabi, 2004; Casson and Lee, 2011; Fligstein and Dauter, 2007). This conflation is compounded by a historical consequence whereby the abstract manifestations of the market promulgated by capitalist discourse (Mackenzie et al., 2007) are derived from markets as an actualized place. As a number of historians have documented (Braudel, 1982; Calabi, 2004; Pirenne, 2014 [1925]) marketplaces played a crucial role in the development and emergence of modern capitalism.

The opportunity for trade may be their *raison d'être*, yet this is not what renders markets an important urban site. After all, the exchange function of markets is hardly exceptional in a global political economy dominated by the hegemony of the market. As Slater and Tonkiss recognize, 'A marketplace... is never simply a meeting of buyers and sellers. Being an embodied event, it always has a specific cultural character, and involves a multitude of social actions and relations. The social density and richness of the marketplace unfolds on a number of levels' (2001, p. 10). Markets produce cultural, social and knowledge capital. They can be iterated as a site of/for consumption; a location of the everyday; an expression of cultural 'authenticity'; a public space; and much more. Jean-Christophe Agnew (1986) argues that it is this very lack of ontological clarity that led to the 'naturalization' of market processes and their infiltration into all aspects of culture and society. 'As a threshold of exchange, the market drew on earlier rituals of passage to distance itself from the many worlds that were indiscriminately mixed within it' (Agnew, 1986, p. 25), and from these multiple ontologies, of which one was place, the exchange aspect of markets was extricated and launched on its own trajectory, leaving a heterogeneous remainder.

Unless otherwise specified, I am referring to an actualized place and the 'embodied event' (Slater and Tonkiss, 2011, p. 10) when I say 'market' or 'markets' in this book. They may also be referred to as a 'marketplace', or sometimes a variation upon this, such as 'bazaar'. Even so, it seems hardly sufficient to say that markets are a place, or that they are emplaced. J. Nicholas Entrikin positions our relationships with place as more than spatial and material. They are ontological and phenomenological as well. Entrikin writes, 'Place presents itself to us as a condition of human experience. As agents in the world we are always "in place", much as we are always "in culture". For this reason our relations to place and culture become elements in the construction of our individual and collective identities' (1990, p. 1). This suggests that discussion about 'What is a market?' – to which one answer is, 'it is a place' – might also need to address 'What is the experience of place in the market?' It is through studying the phenomenology of place in markets that we can come closer to identifying or locating the distinctive qualities of markets.

Making place in markets

'The study of the city is the study of what things emerge in the city', writes Peter Langer (1984, p. 99). I want to apply this formulation to the study of markets, and one of the things to emerge from markets is place. Place is a crucial co-efficient for the markets that I am talking about here – as is acknowledged in that word, market*place*. As Aspers points out, the connections between markets and place are 'observed not only if we trace the phenomenon… but also in its Latin etymology, *mercatus*, which refers to trade, but also place' (2011, p. 4). Agnew (1986) also identifies an interdependence between market and place that has material, located dimensions.

Certainly, place has spatial dimensions, but it also emerges phenomenologically from the space of the market through making. The understanding of making that I am using here comes from Tim Ingold:

> Making, then, is a process of correspondence: not the imposition of preconceived form on raw material substance, but the drawing out or bringing forth of potentials immanent in a world of becoming. In the phenomenal world, every material is such a becoming, one path or trajectory through a maze of trajectories.
>
> (Ingold, 2011, p. 31)

When Ingold talks about making, he says 'we learn by doing' (2011, p. 13). Doing is the phenomenological counterpart to the epistemological 'knowing from the inside: a correspondence between mindful attention and lively materials' (Ingold, 2011, p. 11). Much of what I know and what I have learnt about markets is through doing: buying, selling, looking, chatting, eating, touching, smelling, moving about, meeting up, hanging around, observing, documenting, taking photographs. This doing is a form of making *in* the market, but it is also the making *of* the market, and making that happens *through* and *with* the market. What I am making is different (but not separate) to other making in, through and with the market. Through correspondence and entanglement, making is infinitely generative, as Ingold explains,

> I want to think of making, instead, as a process of growth. This is to place the maker from the outset as a participant in amongst a world of active materials. These materials are what he has to work with, and in the process of making he 'joins forces' with them, bringing them together

or splitting them apart, synthesizing and distilling, in anticipation of what might emerge. The maker's ambitions, in this understanding, are altogether more humble than those implied by the hylomorphic model. Far from standing aloof, imposing his designs on a world that is ready and waiting to receive them, the most he can do is to intervene in worldly processes that are already going on.

(Ingold, 2011, p. 22)

This distinction that Ingold draws between making as 'a process of growth', where the maker is 'a participant in amongst a world of active materials' and making as imposing design on the world, is the distinction, I would suggest, between *making place* and *placemaking*. The former, in the context of markets, is an open-ended process of becoming (Deleuze and Guattari, 1987), where place in the market is made through ongoing participation, intervention and improvisation in, and with, the materiality and phenomena of the market. These are micro-processes, intimate in scale, involving close relationships between senses, bodies, space and materials: the situating and setting up of a stall, the arranging of merchandise, the feel of something as we pick it up to examine it, the limited or extended exchange between vendor and (potential) customer, the movement and the path of our bodies as we negotiate the space of the market. Place here is emergent. Placemaking, on the other hand, is an imposition, a model where space and the social uses to which it is put are manipulated or shaped to produce place. In other words, place is a product.

Atmosphere and markets

In the basement of Market City, a 1990s shopping centre hulking behind the residual façade of Sydney's former central produce market, is Paddy's Market. Paddy's has been operating in Sydney, in one form or another, for almost two centuries. The concrete floors and walls, exposed pipes, and 'clearance height' signs in the space where it is now located communicate the aesthetics and practicalities of a parking station. It is hard to conceive of a market space more antithetical to the lofty iron and glass pavilions of the nineteenth- and twentieth-century market halls still standing in many cities around the world, or to the conviviality of an urban street market (Mehta, 2013). Michael Christie reports that the inauspicious situating of the current Paddy's was not popular when first proposed as part of redevelopment plans of its former site, yet anticipated that even though 'critics will again bemoan the lack of atmosphere,... that is something that [the market] will recreate in time, irrespective of the space it occupies' (1988, p. 149).

Christie's prediction was correct. When you visit the basement today, the atmosphere is lively and sensorially rich. Sections of the walls have been covered with bright green paint and red lanterns have been strung up between stalls, but the contribution of these basic concessions to aesthetics is minimal. Through the correspondences generated between senses, bodies and materials, making in, with and through the market has transformed an unwelcoming and functionalist space into place.

Sights, smells, sounds, tastes and touch are crucial to the experience of markets. Luce Giard captures the phenomenological link here:

> The visit to the market was the time for a marvellous gestural ballet, for winks and funny faces: the outstretched index finger lightly touched the flesh of fruits to determine their degree of ripeness, the thumb tested the firmness of the radishes, a circumspect glance detected the presence of

Figure 1.1   Paddy's Market, Sydney

bruises on the apples, one smelled the scent of melons at length as well as the odor of chevre cheeses, one muttered comments about the relationship between quality and price.

(de Certeau et al., 1998, p. 205)

For this reason, many mimetic, artistic and discursive representations of markets are construed through the senses. They range from Emile Zola's nineteenth-century literary depiction of Les Halles –

Florent kept bumping against hundreds of obstacles – porters taking up their loads, saleswomen arguing in loud voices. He slipped on the thick bed of stumps and peelings that covered the footpath and was almost suffocated by the smell. At last he halted, in a sort of confused stupor, and surrendered to the pushing and insults of the crowd; he was nothing but a piece of flotsam tossed about by the incoming tide.

(Zola, 2007 [1883], p. 30)

– to quotidian experiences reported in a British food magazine:

'I still go to Waitrose or Sainsbury's [supermarkets] for basics and canned goods, but I wouldn't think of buying fresh food there,' she says. 'I love to touch and smell the food in the market, and then go home and cook up something seasonal; there's nothing better.'

(Baldwin, 2011, p. 53)

Place and sense go hand-in-hand, as Steven Feld recognizes: 'As place is sensed, senses are placed; as places make sense, senses make place' (1996, p. 91). It is through the interrelation between place and sense that atmospheres in markets come into being. Christian Norberg-Schulz's (1980) theories on the phenomenology of place were based on a typology of constitutive elements for *genius loci*. *Genius loci* is a concept that Norberg-Schulz 'described as representing the sense people have of a place, understood as the sum of all physical as well as symbolic values in nature and the human environment' (Jivén and Larkham, 2003, p. 70). According to Norberg-Schulz's typology, atmosphere is 'the most comprehensive property of any place' (1980, p. 11).

To talk about the atmosphere of a market is almost a cliché. It is nevertheless something that many would agree markets have, a perhaps ineffable but nevertheless distinctive attribute experienced most fully when one is emplaced in a market. This is confirmed by Irène, one of the participants in Pierre Mayol's study of everyday practices of eating, cooking and shopping in the city of Lyon (de Certeau et al., 1998). Irène describes her sensory responses to her regular market in terms of atmosphere. She comments that the atmosphere of markets distinguishes them from other provisioning places and spaces.

Irène: When I go shopping, I go to the rue Saint-Louis or the big market on boulevard Davila, where I sometimes go on Sundays. That's a real market. I really like that a lot!

Marie: It's a pleasure, the market, yes…

Irène: Yes, it's really delicious! […]

Marie: I think the market is really a pleasure for the eyes.

Irène: Yes… and also a pleasure for other reasons: there are the smells, and then the atmosphere, and the choice too. You have ten kinds of potatoes all next to each other, you have a lot of greens, there's a sort of marvellous abundance. It's not like a big supermarket where every thing is spread out and offered in a provocative way; at the market, it's proposed in a much more natural way, much less ostentatious, you see. To get you to buy, the people at the market have to make a much greater effort, they have to call you over, and shout and make an unbelievable sales pitch, whereas at Monoprix, everything is offered just as it is, immobile, without movement. However, I find Monoprix to be very practical to do your shopping very quickly.

(de Certeau et al., 1998, p. 240)

In the conversation above, Marie also apprehends the market sensorially, commenting that 'the market is really a pleasure for the eyes'. Irène agrees and extends the sensory spectrum: 'Yes… and also a pleasure for other reasons: there are the smells…' To her inventory of pleasures, she quickly adds 'atmosphere', which she fleshes out with visual, that is, sensory detail such as colour, but also with material qualities such as abundance. We might surmise that Irène's perception of atmosphere is a sensory response to the materiality of the market, but what she is also getting at here is 'the ambivalent status of atmospheres. On the one hand, atmospheres are real phenomena. They "envelop" and thus press… "from all sides" with a certain force. On the other, they are not necessarily sensible phenomena' (Anderson, 2009, p. 78). In Irène's narrative, she perceives the market's atmosphere to be both part of and more than her sensory experiences of the market.

Norberg-Schulz, like other theorists of atmosphere such as Gernot Böhme (1993, 2013) and Tonnino Griffero (2014), talks about atmosphere as an aesthetic and emotional phenomenon. Sarah Pink and Kerstin Leder Mackley extend Norberg-Schulz's theories by suggesting that atmospheres are place-based sensory perceptions of 'the encounters between people, materials and other elements of the environments of which they are part' (2014, p. 6). Jean-François Augoyard has also observed that atmospheres are never divorced from material conditions, and that 'side by side with perception, the significance of individual, social, cultural, economic factors projects contextuality' (1998, p. 7). Importantly, Augoyard stresses that atmospheres are emplaced in the immediate environment and thus dependent on 'spatiotemporal incorporation of the morphological and material qualities of place' (1998, p. 7).

I suspect that we often have the relationship between atmosphere and place wrongly figured. We frequently ascribe the production of atmospheres to place. Place does not produce atmosphere, because atmosphere is not a product. Pink and Mackley draw on Ingold when they say that 'atmospheres are emergent from processes of making... Atmospheres are not as such products but they are *produced* or *emergent* ongoingly as people improvise their ways through the world' (2014, p. 6, emphasis in the original). The making of atmospheres therefore corresponds and/or emerges concurrently with making that produces place.

This might account for why placemaking can fail in its attempts to create atmosphere, because it treats place and atmosphere as end-products. Planned, manipulated or enforced atmospheres often don't work because atmospheres are as open-ended and incomplete as the processes of making from which they emerge. Ben Anderson elaborates:

> atmospheres are unfinished because of their constitutive openness to being taken up in experience... They are indeterminate with regard to the distinction between the subjective and objective. They mix together narrative and signifying elements and non-narrative and asignifying elements. And they are impersonal in that they belong to collective situations and yet can be felt as intensely personal.
>
> (Anderson, 2009, p. 8)

Anderson's understanding of atmospheres is influenced by Gilles Deleuze and Félix Guattari's (1987) notion of becoming. Becoming, as you might remember from a passage quoted earlier, was another way that Ingold talked about making (2011, p. 30). Clifton Evers (2014) also draws on becoming to understand the making of atmospheres in a Shanghai night-market.

Another way that Evers thinks about becoming and atmospheres is through the work of Situationist Abdallah Khatib. In 1958, Khatib psycho-geographically mapped Paris' legendary central markets, Les Halles. Khatib cannot have been oblivious to Les Halles' symbolic and material importance to Paris and surely this is part of what drew him to remake the map in the first place. In Khatib's cartography, the flows of the capitalist city act as 'logjams' and 'barricades' to making in Les Halles. Khatib's psycho-geography of Les Halles deployed atmosphere (*ambiance*) as a way of experiencing urban place and space anew, free from the striations of hegemonic spatializations:

> The architecture of the streets, and the changing décor which enriches them every night, can give the impression that Les Halles is a quarter that is difficult to penetrate. It is true that during the period of nocturnal activity the logjam of lorries, the barricades of panniers, the movement of workers with their mechanical or hand barrows, prevents access to

cars and almost constantly obliges the pedestrian to alter his route (thus enormously favoring the circular anti-dérive).

(Khatib, 1958, n.p.)

Markets and cities

Les Halles were was immortalized as the 'Belly of Paris' in Emile Zola's 1883 social realist novel *Au ventre de Paris*, and subsequently become a trope for the centrality of the market (as political economy) in the city of modernity (Mead, 2012; Parham, 2015; Wakeman, 2007). The erection of architect Victor Baltard's nineteenth-century temples to capitalism at Les Halles also functions as a metonym for modern urbanism (Mead, 2012; Wakeman, 2007), as does their demolition in the 1970s to make way for yet another urban renewal project. The issues that affect markets, ranging from environmental sustainability and food security, to urban renewal and access to public space, and cultural heritage and tourism, are at the core of national and international debates on urban design, planning and policy. Informal and formal markets continue to exist and to emerge in urban areas because they provide spatial, social and material infrastructure for those who live and work in cities. They provide many and varied benefits to city life – employment, skills training, community, belonging, cultural exchange and innovation, urban regeneration, tourism, and recycling are only some of them (Evers and Seale, 2014).

In the 1970s, Sydney's wholesale produce market was relocated from the centre of the city to suburban Flemington. Almost 40 years later, a market worker at the Flemington site told a film crew: 'Sydney Markets: it's a city within a city' (Murphy, 2015). The market may have moved to suburban Sydney, but the link between market and city endures in the collective memory, and is expressed through the metaphor of city as market and/or vice versa. One of the images in Peter Langer's typology of 'the image of the city as organized diversity' (1984, p. 99), which he argues is a dominant way of conceptualizing the city in sociology, is the image of the city as bazaar. Basing his reading on Georg Simmel, Langer states that 'Market imagery refers not to the actual marketplace but to the social richness, the equivalent of a bazaar or fair' (1984, p. 102). The city as a socially heterogeneous space is therefore metaphorized as the socially heterogeneous space of the market. Langer metaphorizes once again when he says that the image of the city as bazaar is conceived largely in relation to network capitalism and is thus a metaphor for the market in its many manifestations in the city. This metaphorization 'imagines the city as a place of astonishing richness of activity unparalleled in nonurban areas. It is a market, a fair, a place of almost infinite exploration and opportunity, a centre of exchange' (Langer, 1984, p. 100).

The metaphor is based on an empirical, historical intertwining of markets and the urban – first in towns, and then in cities – that is both morphological and conceptual. In Mandarin, the words for 'town' and 'market' are linguistically linked. Bazaars and souks throughout Central Asia, the Middle East and North Africa are a convergence of markets and the urban (Parham, 2015; Tangires, 2008). In the UK, 'market towns' like Ludlow (Watson and Studdert, 2006) developed because they were the location of a successful market, which led to an expansion of the town through morphological processes of 'market concretion' and 'market colonization' (Conzen and Conzen, 2004, p.42). Markets have been central to the secular morphogenesis of cities (Conzen and Conzen, 2004; Conzen and Larkham, 2014), even when overseen by religious authorities (Slater, 2014).

Historian Henri Pirenne cautioned that the presence of a market alone was not sufficient for the development of a city, writing 'any number of places, ... although equipped with a market... never rose to the rank of city' (2014 [1925], p. 88). Geography was also crucial. The geographical factor could be a topographical advantage (Parham, 2015) of the sort described by Casson and Lee in their history of markets: 'A well-situated market – for example, near the bridging point of a river – would stimulate the growth of a town' (2011, p. 14). Or, it could be a relational geography to other markets, as in central place theory (Plattner, 1985). Merchants and traders not only distributed goods as they travelled to urban market hubs along transnational, long-distance trading networks (Casson and Lee, 2011) like the Silk Road, but also transmitted ideas and practices of the urban as well. Whichever way you look at it then – spatially, politically, socially, culturally – the history of the city is entangled with markets (Calabi, 2004; Conzen and Larkham, 2014; Mumford, 1991 [1961]; Pirenne, 2014 [1925]; Tangires, 2008; Wirth, 1938). As Fernand Braudel says in his definition of urban settlements, 'No town is without its market, and there can be no regional or national markets without towns... the town in other words generalizes the market into a widespread phenomenon' (1981, p. 479).

In this book, I understand both markets and cities to be nodes where multiple and proliferating heterogeneous flows are organized and disorganized. Khatib's psycho-geography of Les Halles is an illustration of a node.

> The essential feature of the urbanism of Les Halles is the mobile aspect of pattern of lines of communication, having to do with the different barriers and the temporary constructions which intervene by the hour on the public thoroughfare. The separated zones of ambiances, which remain strongly connected, converge in the one place: the Place des Deux-Ecus and the Bourse du Commerce (Rue de Viarme) complex [*sic*].
>
> (Khatib, 1958, n.p.)

Nodes, as defined by Kevin Lynch in his typology of images of the city (1960, pp. 72–78), are 'concentrations and junctions'. Lynch largely positions nodes as spatial, as breaks, crossroads or meeting points in material urban infrastructure – although he does allow that we may perceive these concentrations and junctions with heightened sensory awareness because of the rupture and/or (re)combination of flows that happens there. Agnew also talks about markets as if they are nodes (although not necessarily urban ones) that were 'often at crossroads or at the borders between villages, ... neutral territory where men might occasionally meet for their mutual benefit' (1986, p. 27). Geographical affordances enhance markets and cities' nodal functions through providing a spatially and/or topographically logical point for the arrangement and rearrangement of flows.

Markets are nodes where material and intangible flows – of people, goods, time, senses, affect – come to rest, terminate, emerge, merge, mutate and/or merely pass through, and are contingent and relational to one another. They, like cities, are nodes that also function as regulating mechanisms to both produce and organize heterogeneous flows of people, goods, capital, food, water, waste, transportation, and so on. In *Life inside the markets*, a documentary series (Murphy, 2015) about Sydney's wholesale fruit and vegetable market, a worker calls this 'organized chaos'.

Those who control the regulatory mechanisms for flows control the polis. In ancient Greece there was no delineation between the agora as marketplace and as the public articulation of political power in the city. Extant material evidence in city centres across Western Europe – including in Antwerp, one of the case studies in this book – demonstrates the propinquity of the marketplace to buildings from which power was exercised. Markets have therefore always been regulated spaces, regardless of their reputation for unpredictability, mess and disorder. In medieval and early modern times, markets were the place for public shaming and punishment. The ostensible other of this public enactment of discipline was the unruly energy of the festival, a close counterpart of markets. But as Bakhtin (1984) has told us, the carnivalesque of the festival was really a licensed affair, where the temporary inversion of the social order merely reinforced the status quo for the rest of the year. Festival and pillory served the same purpose – the disciplining of the public and of public space in the city.

Clearly, then, there are thresholds established around and within the space of a market, and with regard to its practices and functions as place. Yet these thresholds do not always define markets as a place of prohibition, sacralization or exclusion. These thresholds were designed to be transgressed through the invitation of others into the marketplace, and therefore the city, as Agnew explains:

> Alien merchants were called *metics*, a title from the root word for intermediary and, generally, for change. The merchants' ambivalent social status corresponded precisely to the kind of socially constructed space that had been interiorized with the incorporation of the marketplace into the civic center. This was an interstitial space, a space "between" inhabited by "go-betweens"; it preserved the affective and cognitive associations of a boundary, a border, a frontier, or a limit. But it was at the same time a culturally inscribed limit, that, by its very ritual and mythological associations, was identified with, if not defined by, its own transgression.

The ancient marketplace was, as the etymology of "limitation" suggests, a *limen*, a threshold.

(Agnew, 1986, pp. 22–23)

This means that even when markets moved from 'the periphery of settlement, in neutral zones or marches lying between villages, tribes, and societies' (Agnew, 1986, p. 20) to the spatial and cultural centre of the city, they maintained structural and substantive liminality.

The limits of the marketplace are ambivalent. Markets are therefore porous spaces in which exclusion, order and discipline are part of a network of heterogeneous flows in which its opposites – inclusion, disorder and transgression – are also present. This porosity points to a metonymical link between market and city. Rebecca Solnit writes that 'Cities had a kind of porousness… cities were impossible to seal against artists, activists, dissidents and the poor' (Solnit and Schwartzenberg, 2002, p. 82). Solnit posits this porousness as a historical condition, which is under threat in contemporary cities due to the restrictions imposed upon urban space by neo-liberal corporate and state agendas. The types of markets that facilitate porousness – informal, second-hand, public – are also frequently under threat of being closed down, relocated or redeveloped, while others where the spatial and social boundaries are less permeable – festival marketplaces, maker and vintage fairs, organic and farmers' markets – are on the rise. Place that emerges from everyday making in unremarkable markets is being superseded by placemaking in specialist markets that supports the mixophobic monoculturalism of gentrification (Bauman, 2011; Solnit and Schwartzenberg, 2002), the displacements of urban renewal, and the devaluing and privatization of public urban space (Low and Smith, 2005).

This premise – that the fate of place in markets and the fate of place in cities form a metonymy – is the subject of Chapter 2. Chapter 2 and the chapters that follow are a series of essays about the constellation of market, place and city, and the conditions and effects of transformation, continuity, disappearance and emergence in this constellation. This is considered in the context of place-image, unremarkable places, displacement, the right to the city and urbanism, and with reference to grounded case studies. In selecting the case studies and contexts for this book, I am mindful of Giorgio Agamben's notion of example as 'a concept that escapes the antinomy of the universal and the particular [that] has long been familiar to us' (1993, p. 8). Indeed, Agamben's formula may be a way that we can think about markets and cities themselves, dialectically moving between the universal and particular. The examples here elucidate the shared and individual attributes of markets as places in cities. Sometimes there is overlap, but just as often these markets, places and cities are worlds apart – spatially, socially, materially and phenomenologically.

2 Markets as metonyms of urban transformation

East London

Metonymy, explain Lakoff and Johnson, 'has primarily a referential function, that is, it allows us to use one entity to stand for another' (2003, p. 36). In this chapter, I want to propose that East London's markets operate metonymically in relation to urban transformation, and more specifically, to the regeneration of East London as a globally visible cultural quarter (Miles, 2010). Following this pattern, transformation of place in markets is metonymically linked to transformation of place. In East London neighbourhoods where working-class and migrant communities relied heavily on the street market for goods and employment, local markets have moved from being traditional retail markets to specialist markets as a response to spatialized discourses of post-industrial urban renewal, gentrification and the Creative City (Mould, 2015; *pace* Mould I am adopting the capital 'C' for this term to denote the discourse, as well as the space). These ideas are investigated with reference to two street markets in the East London borough of Hackney. Ridley Road is a lively everyday market in a gentrifying neighbourhood that is 'notable for its ability to cater to the tastes and budgets of a wide range of the city's most marginalised migrant groups' (Rhys-Taylor, 2014, p. 45; see also Rhys-Taylor, 2013; Watson and Studdert, 2006; Wessendorf, 2013, 2014). Broadway Market, on the other hand, is a 'boutique' market, which produces and captures the place-image (Shields, 1991) of the 'new' East London through its emphasis on artisanal, handmade and vintage products, and its alignment with the creative industries (Parham, 2012; Taylor, 2011).

'The story of London is the story of its markets'

I first started thinking about metonymies of place in markets and cities because I was interested in contemporary British writer Iain Sinclair's treatment of place in East London. Sinclair's sustained literary engagement with the spatial, political, cultural, social, mnemonic and sensory dimensions of place in East London over four decades of writing constitutes a singular archive of urban transformation in the city's east from the 1970s on (Seale, 2008). In his writing, photography and filmmaking, Sinclair uses documentary modes of social research, reportage and interview. His narratives are fuelled by the observed details of everyday practices and encounters that communicate place. Sinclair has cited the street market as an influence for his gravitation towards East London as a place to write about, and from which to write: 'Here was my raw material, a job for life, picking at a mythology of place: subterranean conspiracies, lost writers, the action in street markets' (2009, n.p.). One of Sinclair's aphorisms, 'The story of London is the story of its markets' (2006, n.p.), resonated for me. Sinclair's metonymy of market and city is more than a recognition of the existential intertwining of the two. It is an acknowledgment that markets are barometers of place, that the place that is being made in a market can be used as a gauge or measure of the place that is being made in the urban landscape in which it is situated. In this chapter, and to an extent in this book, I am working with a slightly modified version of Sinclair's formulation: the story of place in a market is the story of place in the city.

Sinclair's first novel, *White Chappell, scarlet tracings* (1987), is an investigation of the mythopoeia of place that is set in East London and in the sub-cultural milieu of the second-hand book trade. The book is based on autobiographical experience. In the 1970s and 1980s, Sinclair had a market stall at Camden Passage in Islington selling second-hand books. During his years as a dealer, flea markets like Cheshire Street off Brick Lane were a source for stock. In *White Chappell*, the book-dealer/rag-pickers are an unfortunate, blighted lot barely existing on the leftovers of others, and Dickensian allusion is satirically applied to accentuate the comic pathos and degradation of their lives:

> Dryfeld growls through the vans, pokes into sacks, storms among the sheds of rag pickers, elbows over terminal waste-lots, where old bones have been spread out to dry, more for exhibition than with any serious expectation of a sale. He snarls back at the caged animals, bird yelp, rancid fish tanks, heavy jaw'd fighting beasts dealt, as they have been for over a hundred years, under the railway arches. The sentiment of the local inhabitants flattered by having some creature whose existence is even worse than their own.
>
> (Sinclair, 1995, p. 38)

Similarly, in a 1997 non-fiction essay, Sinclair's prose is infinitely inventive in characterizing the goods at the now obsolete Farringdon Road second-hand book market as refuse:

> George had, over the years, dispersed acres of country house libraries, … remorseless tides of salvage. Rare Victorian pamphlets, plump Edwardian bindings, railway fiction – he graded the lot, hemp sack or auction table. He kept the culture of print in flow. He served it like a pest controller, a water bailiff. Perched above the Fleet ditch, he shovelled the failed remnants, the picked-over dross, into the corporation's dustcarts.

These Farringdon Road barrows were the court of final appeal. After the frantic ceremonies of the predators there was extinction.

(Sinclair, 1997, p. 19)

As hyperbolic as Sinclair's portraits of East London markets and their communities are, there is a recognition here that the heterogeneous mix of identities, encounters, practices and goods (Stallybrass and White, 1986) enables heterotopic potential and a Lefebvrian right to the city (Lefebvre, 1996). The heterogeneity of these markets is metonymical to the heterogeneity that is considered by many to be a defining attribute of the city (Bauman, 2011; Langer, 1984; Massey et al., 1999; Sennett, 1993 [1974]; Wirth, 1938; Young, 1990). These, however, are not the type of markets that are used in placemaking and place-marketing for East London today. Cheshire Street's junk markets have been cleaned up and rebranded as 'vintage', and Farringdon Road closed up shop in 1994.

Metonymies of markets and city

Markets have been prevalent throughout London since early modern times due to what Donatella Calabi calls a diffuse model whereby 'the public spaces used by those who participated in the urban market were spread out over a fairly vast area of the city centre' (2004, p. 91). This is, Calabi says, a different model to the centralized 'market square' model in other towns and cities in the UK and Europe, and the intensely centralized model I write about in Sydney in Chapter 5. This morphological circumstance may explain why street markets have long been synonymous with London, and particularly East London, in the popular imagination. Certainly, the perception of East London street markets has been metonymical to the historical perception of East London itself – the menacing, unknown, exotic, dirty 'other' to the London of political and financial power and the London on tourist postcards (Jacobs, 1996). Many of the words associated with the East London market – cheap-jack, peddler, duffer, hawker, huckster – insinuate illegal, or at least informal activity. Henry Mayhew, in his groundbreaking multi-volume sociological study of London's working and underclasses, *London labour and the London poor* (1861), revealed the street market to be the stage for capitalist micro-commerce of a vast and widely distributed scale that supported large numbers of vendors and consumers. Mayhew documented his visit to Petticoat Lane, an informal market in East London:

You meet one man who says mysteriously, and rather bluntly, 'Buy a good knife, governor.' His tone is remarkable, and if it attract attention, he may hint that he has smuggled goods which he must sell anyhow… Another man, carrying perhaps a sponge in his hand, and well-dressed, asks you, in a subdued voice, if you want a good razor, as if he almost suspected that you meditated suicide, and were looking out for the means! This is another ruse to introduce smuggled (or 'duffer's') goods…

These things, however, are but the accompaniments of the main traffic. But as such things accompany all traffic, not on a small scale, and may be found in almost every metropolitan thoroughfare, where the police are not required, by the householders, to interfere.

(Mayhew, 1861, p. 44)

In the 1930s, Mary Benedetta travelled to Petticoat Lane and had a different experience buying a knife. Contrary to the market's reputation, and in spite of being told off by one stall-holder for not making a purchase, Benedetta was pleasantly surprised at the quality and price of merchandise, reporting, 'Here is a real pantomime of life. Some of the quaintest personalities trade in this market, and eighty per cent of the wares are said to be genuine' (Benedetta and Moholy-Nagy, 1972 [1936], p.1). Benedetta's travel narrative accompanied a series of photographs of London street markets by legendary Bauhaus photographer László Moholy-Nagy. Moholy-Nagy took his camera out into the street to capture the 'real' London, which he believed was distilled in its markets. In his foreword, Moholy-Nagy was motivated, as Mayhew was, to dispel the image of the street market as the site of solely nefarious activity:

To many people's minds the street market still suggests romantic notions of showmen, unorganized trade, bargains and the sale of stolen goods... in my opinion these markets are primarily to be regarded as a social necessity, the shopping-centre, in fact, for a large part of the working-class.

(Benedetta and Moholy-Nagy, 1972 [1936], p. vii)

Metonymies of market and city are evident in iterations of place in contemporary East London too. In the 1970s, middle-class gentrifiers started moving into Spitalfields, ostensibly to 'save' the area's Georgian heritage, which was under threat from neglect, urban renewal and adaptive reuse for the garment trade (Benton, 1985; Jacobs, 1996). However, enthusiasm for local heritage did not extend to Spitalfields Market, a wholesale fruit and vegetable market that had been in the neighbourhood since the seventeenth century. The market was a metonym of an East London that newcomers hoped to relegate to the past – unsanitary, uncivilized, incommodious. They lamented its litany of inconveniences including noise, crowds, refuse, vermin and congested roads (Benton, 1985). The wholesale market was relocated to Leyton in 1991, and while the future of its former site was determined, an antiques, handicraft and clothing market that was a better fit with the tastes and practices of the new residents was set up. Ultimately, the site was redesigned by Sir Norman Foster, and a multistorey complex with office space, upscale eateries, boutiques for global brands such as Camper and Fred Perry, and a downsized marketplace in the heritage-listed Horner Building was completed in 2005. In a curious inversion, the market at Leyton, which continues the wholesale produce trade, is called 'New Spitalfields Market', whereas the reconfigured original marketplace, which is nothing like the former market, has been christened 'Old Spitalfields Market'.

Figure 2.1   Hanging out at Broadway Market

Another metonym of place in the 'new' East London is the Kingsland Waste market, an increasingly marginalized flea market where waste and commodities mix promiscuously. The Waste's second-hand stalls barely exist on the edge of Kingsland Road, where they are under constant scrutiny from the local authorities who claim concerns about health and safety and intellectual property rights as a means of regulating the market. The market's real offence is that its place-image of mess, disorder and informal commerce is contrary to current official place-images of East London as Creative City.

The battle for Broadway Market

One metonym that *is* consistent with East London's contemporary place-image as Creative City (McRobbie, 2004; Pratt, 2011) is Broadway Market. Since 2004, when it was reinvented as a specialist market, Broadway Market (which is the name of both the market and the street in which it is held) has been the emergent other to vanishing East London markets. In the 1930s when Benedetta was writing, the borough of Hackney had seven street markets with 'just under a thousand stalls' (Benedetta and Moholy-Nagy, 1972 [1936], p. 140). Of the seven, only Ridley Road and Well Street have retained their original purpose as everyday or traditional retail markets. In a 2009 document, Hackney Council explicitly linked transformation of the borough's markets to urban development and transformation in the area; specifically, construction of infrastructure for the London 2012 Summer Olympic Games. The council stated:

> We know that the East London line extension will be coming to Shoreditch and Dalston and wider plans are being drawn up to regenerate these areas. This is an exciting time for markets with investment opportunities and a revival of shopper's interest.
>
> The Market Service is keen to raise standards so that everyone feels safe and comfortable using our markets and improved market facilities.
>
> (Hackney Council, 2009, p. 4)

Hackney Council's focus on its markets as sites of, and for, regeneration is substantiated in their 2010 cultural policy framework, which links local markets to creative economy-driven urban renewal in the area (Hackney Council, 2010). The council's stated policies are intriguing because in the years prior, the council was actively obstructing Broadway Market's evolution into a specialist market. As Patrick Wright (2009) recounts, the current Saturday market was an initiative that originated with shop-owners and traders along Broadway Market in what Susan Parham has referred to as 'renewal from below' (2012, p. 154). This plan to replace a desultory street market with a revitalized version faced opposition from Hackney Council and from developers to whom the council had sold off commercial properties along the street because it did not fit with their own development plans. Local writer Hari Kunzru, who followed the story for the national *Guardian* newspaper, discovered that tensions over place in the market were not about gentrifying newcomers versus long-time residents, but about privatization of public space and 'the connection between interesting cheeses, international finance, local government impotence and the impending tsunami of upward-mobility that is the 2012 Olympic games' (2005, n.p.). Kunzru noted that 'the result of the council's scrabble to clear its debt has been to put the vast majority of this little street, not into the hands of the people who live and work in it, but of just three big property developers... Welcome to the globalized Hackney street market' (n.p.). Traders and residents, regardless of how long they had been in the area, employed the rhetoric of battle in their resistance to externally produced discourses and practices of revitalization, and the implication that what was already in Broadway Market was valueless (Smith, 1996).

The battle for Broadway Market mobilized complex ideas of belonging. Newcomers who had a level of self-reflexivity about the implications of their presence were sufficiently concerned about the social plurality of the area and the retention of public space to be actively involved in protests and occupations of buildings along Broadway Market following the privatization. A documentary, *The battle for Broadway Market* (James, 2009), recorded the long campaign by locals against rent rises and the evictions of businesses that had been in the street for decades.

In the video, the third-generation proprietor of the local pie and mash shop compared Broadway Market in the 1990s to a bombed-out Beirut. Wright's history of the market also concedes that there was a need for improvements to existing infrastructure. The protesters were therefore united in demanding an alternative, locally produced urban regeneration, not one dictated from without and/or above. As an example of this, Broadway Market has been successful. After difficult beginnings, the market flourished, as Susan Parham writes:

> [T]he revitalized food market rapidly became very successful socially and economically among both a local and wider visitor catchment, while the redevelopment of the market and surrounding quarter likewise demonstrated some areas of contestation… There was, though, more unanimity about the importance of the market to the area's regeneration and the considerable social mix that the market catered for. One of those interviewed, a food trader, agreed that the food market had led the regeneration of the area and pointed out that residential incomers had been part of what had made Broadway Market sustainable in economic terms. A number of interviewees suggested that the Broadway area was enjoying a mix of incomers and existing residents.
>
> (Parham, 2012, pp.154–158)

This is confirmed in a video on YouTube (Love, 2010) about the market, where interviewees who could be identified visibly as hipsters or 'arty' (Parham, 2012) deployed the rhetoric of place, the local and belonging to describe their affinity for the market. Another of Parham's interviewees was more blunt: 'I see the market as gentrifying space' (2012, p. 160). Stating that the market is catering to existing local residents is true, but it is more an acknowledgment of advanced gentrification in the area, a process for which the market is a vehicle and symbol (Wessendorf, 2014), than an acknowledgment that the market is meeting the needs of all local residents.

Creative destruction in the East London market

The artisanal and organic produce, vintage clothing and handmade crafts on sale at Broadway Market are not about necessity or custom, but function as markers of distinction, cultural competence and the accumulation of cultural capital (Bourdieu, 1984). On a Saturday afternoon in the summer of 2013, some of the things shoppers strolling the market could purchase were a flat white coffee advertised as 'Australian'; vintage vinyl records; Vietnamese summer rolls; handmade kids' clothes made out of screen-printed fabric; hot salt beef on a Brick Lane bagel; and a 1980s pink floral jump-suit from a rack of second-hand clothes. These types of products are synonymous with the consumption habits of those whose work and leisure comes under the umbrella of the creative industries. In an article published in Jamie Oliver's eponymous magazine *Jamie* (Baldwin, 2011), several vendors at Broadway Market repeated a narrative that is common in creative industries discourse whereby they had left jobs in other sectors to pursue their 'creative dreams', or they were working on their day off to do what they 'love' (Seale, 2012).

> 'Everyone here is a massive foodie,' says Tamami Haga, a mother of two who sells her homemade cakes at Broadway every weekend. She's a fairly typical stallholder, an independent producer who works one day a week doing something she loves (she even blogs about it…). 'It's all about the people,' she says. 'I love interacting with customers and making friends with them.' Tamami is up at 6am every Friday to begin baking her berry tarts and rich chocolate brownies, and doesn't finish until early Saturday morning – mere hours before her stall opens…
>
> Since leaving finance three years ago for her first love – the 'natural fusion' of Vietnamese cuisine – Van Tran has been serving Broadway Market, developing close relationships with customers and colleagues alike. It has, she says, a unique sense of community…
>
> Since leaving a career in menswear design, Soli Zardosht has devoted herself to what she calls 'a true labour of love'…
>
> Each weekend Sabrina and Sol Negron leave careers in education and architecture, and recreate Mexico in the corner of a schoolyard-turned-market.
>
> (Baldwin, 2011, pp. 52–61)

With these types of narratives dominating in media images of Broadway Market, it is no surprise that it has become shorthand for the spatial colonization of London's East by the creative class (Florida, 2002). Broadway Market has become a convenient lightening rod for backlash against creative workers (practitioners) and hipsters (consumers). A satirical Tumblr blog calling itself Hackney Hipster Hate parodied the social demographic at the market (and in East London more generally) in a photograph of two 'Hackney Hipsters' with the following speech bubble exchange:

First hipster: You know you should totally join our Broadway Market residents committee.
Second hipster: That would be really cool, but I've only lived here for three days…
First hipster: Yeah me too.

> (Hackney Hipster Hate, 2010)

The transformations in and near Broadway Market act out a form of creative destruction, Karl Marx's theory that capitalism was predisposed to destroying not only alternative economic systems but, through an act of cannibalization or destruction, annihilating itself. In recapitulating Marx, Marshall Berman could be describing what has happened to East London as a result of ongoing urban renewal, and especially in the lead-up and aftermath to London 2012:

> The truth of the matter, as Marx sees, is that everything that bourgeois society builds is built to be torn down. "All that is solid" – from… the houses and neighbourhoods the workers live in, to the firms and corporations that exploit the workers, to the towns and cities and whole regions and even nations that embrace them all – all these are made to be broken tomorrow, smashed or shredded or pulverized or dissolved, so they can be recycled or replaced next week, and the whole process can go on again and again, hopefully forever, in ever more profitable forms.
>
> (Berman, 1983, p. 99)

It is tempting to read Broadway Market, as Sinclair does, as the all-too-visible counterpart to the now-vanished Hackney Wick flea market, which fell victim to creative destruction when its terrain was swallowed up by the Olympics site.

> Hackney Wick disappears into a pre-Olympic limbo of exaggerated promises and present suspension of liberties. But in another part of the borough, Broadway Market, jellied-eel mythology gives way to a pastiched Islington. No 50p tat here: discriminations of olive oil, fancy breads and a stall selling lush volumes by notable photographers.
>
> (Sinclair, 2006, n.p.)

Urban renewal as a mode of creative destruction takes on a certain grim irony when it uses the Creative City as a 'vehicular idea' (Peck, 2012). This irony has not been lost on Sinclair, who has communicated ambivalence about his role in the London writing genre, for the reason that he is aware of his double role as both commentator on and agent of urban transformation in the city's east. In writing about East London as an overlooked territory, Sinclair has contributed, albeit unwillingly, to a process which makes a place visible so that it can become a space for consumption (Zukin, 1995). He has even used gentrification, for which he believes Broadway Market is metonymical, as the impetus for *Hackney: that rose-red empire* (2009), his 600+ page homage to pre-Olympics East London:

> I took a contract, as you do, for a totally different kind of book… Then, one morning, I was going through Broadway Market and I met about 20 people I knew, but from all over London, all buying a loaf of bread and a bag of tomatoes for 20 quid, and I thought: this is it. I've got to start now, or it's gone.
>
> (Cooke, 2009)

Sinclair is one of the 'shock troops' of gentrification (Ley, 2003), that is, the artists who first move into an area. Now, he would be classified as a member of the creative class.

A tale of two markets

At the same time as it is promoting itself as a renewed food quarter (Parham, 2012), Broadway Market's official website evokes nostalgic ideas about previous East London markets through the figure of the barrow boy (albeit all grown up) on its homepage. An image of John, who runs a fruit stall at the market, is accompanied by the following text:

> Barrow boys have been welcoming shoppers to Broadway Market in Hackney since the 1890s… John and his mate Tony… may be the last in the line. John started selling fruit and veg on the market nearly 50 years ago… now his barrows are the centrepiece of the revived Saturday market.
> (Welcome, n.d.)

Exceptionalizing John and Tony as authentic East Londoners only draws attention to the discrepancy between residents from adjacent housing estates and the clientele of the reconceptualized market. One Saturday I sat at the junction where Broadway Market meets Benjamin Close and observed that most of the people coming from nearby Welshpool House used Broadway Market as a social space, stopping to have a chat in the street with each other, but, with the exception of John and Tony's stall, they avoided the market and boutiques in favour of vernacular spaces (Zukin, 2012) such as the Costcutter supermarket, the post office and the betting shop.

James Meek's piece for the *London Review of Books* in the aftermath of the 2011 London riots situated Broadway Market as an exemplary urban space for exhibiting the propinquity of deprivation and affluence in globalized cities. Meek wrote: 'When Broadway Market actually becomes a market on Saturdays it is as if the council-owned tower blocks and estates behind, around and in between the gentrified patches, where less well-off and poor people live, belong to some other dimension' (2011, n.p.). The aforementioned YouTube video about the market (Love, 2010) captures the disjuncture that I observed, and about which Meek wrote. In the video, the market's managers (who are inexplicably dressed up as gangsters as if to underline that they are pretending to be East Londoners) tell a local resident who complains about prices at Broadway Market to go to Ridley Road Market. This exhortation to go elsewhere seems at odds with Broadway Market's statement on community on its website, which claims that 'The market benefits the whole of Hackney' (Community, n.d.).

It is not, however, at odds with an ongoing place-image internally generated by the market as the *not*-Ridley Road. For example, a blog post on the Broadway Market website titled 'Ridley Road rat scandal' decried the lack of local council oversight that allowed the unchecked sale of illegal bush meat at Ridley Road.

> Officials are already moving to protect their backsides.
> Markets Department bosses point out that this is not their responsibility because the meat was found being sold in shops and shops are not

considered to be part of the market. So market inspectors – who are on the street day in day out – could not be blamed for failing to notice the odd rat...

Environmental health officers should be given the resources they need to do their job. And, please, let us have no more excuses.

(Ridley Road rat scandal, 2012)

In calling for increased surveillance, and implying that they had nothing to conceal, Broadway Market's management distanced their market from Ridley Road Market.

The place and the atmosphere at Ridley Road are quite different to Broadway Market because there are different types of making going on here. Making at Broadway Market is energized by the artful purposelessness of *flânerie*, loitering, watching and being seen. Leisure and consumption are mobilized in the formation and expression of identity. It's a day out, a place to catch up with friends. Used to early market starts, I arrived at Broadway Market at the opening time of 9am, but most of the stalls were still setting up and weren't serving food. The market didn't really get going until 11am. Ridley Road, on the other hand, is on every day (except Sundays), and has an advertised start time of 6am. Place and sociality are not derived from eating or hanging out, and there were no readily available formal seating areas to encourage these types of making place.

Compared to Broadway Market, which has an emphasis on snacks and upcycled street food, Ridley Road is about shopping for the larder and the home. The discipline of provisioning, linked as it is to the discipline of keeping house, is matched by the disciplining of public space. Even in the 1930s, Benedetta's visit to Ridley Road was framed by notions of regulation: 'Street-traders are subject to laws and discipline just as any other citizen... The street market inspectors are the guardian angels of all the stall-holders. They are also very strict disciplinarians where licenses are out of date' (Benedetta and Moholy-Nagy, 1972 [1936], p. 135). Compared to Broadway Market, surveillance was more overt. Contrary to the complaints of Broadway Market's management, and to local, national and international media preoccupation with informal, illegal and unhygienic practices at the market (Lynn, 2012), council workers were very visible, particularly those maintaining sanitation.

Ridley Road is an East London palimpsest. As I walk from Dalston Lane to Kingsland Road, I encounter a multi-layered landscape that has been subject to erasure and addition. A 'halal' sign on a shiny food truck selling hot dogs and burgers is juxtaposed with the peeling and fading wall poster advertising a McDonald's around the corner (see Figure 2.2). As I continue walking towards the corner where Birkbeck Road meets the market, I come to the irrepressible bright-orange façade of Mr Bagel. A reference to the days when the market was predominantly Jewish traders (Hackney Council, 2009; Watson, 2009), 'Rice n Spice Caribbean and Indian Food' has now been added to its name. You can

Figure 2.2   Ridley Road Market

still order bagels here, but you can have curries and jerk chicken too. Making
in Ridley Road produces 'cosmopolitan conviviality' (Gilroy, 2004), 'everyday
multiculturalism' (Watson, 2009) and 'commonplace diversity' (Wessendorf,
2013). The foodie cosmopolitanism on display at Broadway Market is a performative exhibition of 'cultural diversity', whereas cosmopolitanism at Ridley
Road is the habitus of 'cultural difference' (Bhabha, 1994). It is not, writes
Rhys-Taylor, 'a unilateral or bilateral exchange between a local host culture and
the marginal culture of migrants' (2013, p. 399), where the issue of 'eating the
other' rears its ugly head (hooks, 1992). Rhys-Taylor concludes that at Ridley

Road Market 'cultural heterogeneity is ubiquitous, conspicuous, but also banal and, pragmatically speaking, unimportant' (2013, p. 403). Ridley Road is Third Space (Bhabha, 1994), a hybrid topography where you can source mangoes, cheap luggage and money transfers.

Mixophilia and mixophobia in the market

In spite of the polarizing gentrification going on around it, Ridley Road has not yet attained what Neil Smith called 'gentrification kitsch, in which cultural difference itself is mass produced' (1996, p. 115). It provides something more than a temporary thrill for middle-class shoppers seeking the street theatre of cultural diversity. In their study of sociality and social inclusion in markets in the UK, Watson and Studdert found that where the socio-demographics of the local community were 'very mixed, such as in Ridley Road, the market appeared to act as a site of mixing and connection in very positive ways' (2006, p. 29), and that 'Ridley Road, of all the market sites, according to traders and shoppers and from all observations, operated as perhaps the most vibrant social space for shoppers and traders between and across many different cultures' (2006, p. 27). Markets like Ridley Road provide a 'cosmopolitan canopy' (Anderson, 2011) where what Sophie Watson calls 'rubbing along' with difference contributes to a sense of place, and 'a form of limited encounter between social subjects where recognition of different others through a glance or gaze, seeing and being seen, sharing embodied spaces, in talk or silence, has the potential to militate against the withdrawal into the self or private realm' (Watson, 2009, 1581).

Peter Langer has written that whether we love or hate the city 'decisively influences one's image of the city. The city may be seen as a place of dirt, disease, crime, pollution, vice, poverty, and other social problems, or it may be seen as a place of culture, art, wealth, employment, vitality, sophistication, and social opportunities' (1984, p. 100). Given the historical, spatial and social intertwining of market and city, I would suggest that people's feelings about markets could work according to a similar schema. The market in Ridley Road is a place whose kaleidoscopic and kinetic heterogeneity is metonymically linked to the heterogeneity of the city, and is therefore as likely to inspire mixophobia as it is mixophilia, as Bauman explains:

> City living is a notoriously ambivalent experience. It repels but also attracts, yet it is the same aspects of city life that, intermittently or simultaneously, attract and repel. The messy variety of the urban environment is a source of fear, yet the same twinkling and glimmering of urban scenery, never short of novelty and surprise, boasts a charm and seductive power that is difficult to resist. Confronting the never-ending and constantly dazzling spectacle of the city is not therefore experienced unambiguously as a curse; nor does sheltering from it feel like an unmixed blessing. The city prompts mixophilia as much as mixophobia.
>
> (Bauman, 2011, pp. 63–64)

The geography of encounter (Valentine, 2008), the possibility of spontaneously meeting the other in the streets is not an attractive proposition for many urban dwellers, even while they welcome it as an abstract notion of urban experience, or within areas demarcated specifically for that purpose. Meek points out, 'Loving the cultural diversity of London as a spectator-inhabitant is not the same as mingling with it' (2011, n.p.). Meek quotes from Slavoj Žižek's 2008 book *Violence*, where Žižek posits this social insularity as essentially neo-liberal in character:

> Today's liberal tolerance towards others, the respect of otherness and openness towards it, is counterpointed by an obsessive fear of harassment. In short, the Other is just fine, but only insofar as his presence is not intrusive, insofar as this Other is not really other ... My duty to be tolerant towards the Other effectively means that I should not get too close to him, intrude on his space... What increasingly emerges as the central human right in late-capitalist society... is a right to remain at a safe distance from others.
>
> (quoted in Meek, 2011, n.p.)

In spite of their reputation for disorder and unpredictability, markets are one of these demarcated spaces. Markets are one of the few urban spaces in the city where people truly feel at ease with difference, whether it be a regulated performance of cultural diversity or autonomous practices of cultural difference. Markets therefore lend themselves to fetishization as representations of urban diversity, and reinforce the mixophobia of the city even as they appear to be promoting mixophilia.

Mixophobia does not solely manifest itself through excluding the group who are subject to the mixophobia. Wessendorf has noted that in Hackney a type of mixophobia emerges, somewhat ironically, from an insistence on mixing, and is directed towards those who are perceived as not demonstrating an 'ethos of mixing'.

> While many people mix across cultural differences in public and associational space, this is rarely translated into private relations. However, this is not perceived as a problem, as long as people adhere to a tacit 'ethos of mixing'. This comes to the fore in relation to groups who are blamed to 'not want to mix' in public and associational space... positive attitudes towards diversity are accompanied by little understanding for groups who are perceived as 'not wanting to mix', a phrase repeatedly used by my informants.
>
> (Wessendorf, 2013, pp. 407–408)

This type of mixophobia extends to Hackney's hipsters who are viewed as leading separate lives in a concatenation of monocultural spaces – one of which is Broadway Market – and hence do not demonstrate the ethos of mixing. Other groups or individuals feel excluded from these spaces, and then self-exclude themselves. Wessendorf (2013) spoke to multiple informants who had lived in Hackney for ten years or longer and who avoided the places where hipsters congregated or the places which had been adapted to hipsters' tastes because they felt they didn't belong. In September 2015, the perception that hipsters were mixophobic generated its own mixophobia in the form of a physical attack on a small business that was deemed emblematically hipster (Khomami and Halliday, 2015).

Broadway Market is arguably one of these contested places; after all, a market had existed previous to the establishment of the current iteration in 2004 (Wright, 2009). It is a metonym of the mixophobic city because it substantiates the monoculture of a gentrifying creative class. 'The drive to a homogeneous, territorially isolated environment may be triggered by mixophobia,' says Bauman, which 'manifests itself in a drive towards islands of similarity and sameness amidst the sea of variety and difference' (2011, p. 64). In urban neighbourhoods which are in the advance stages of gentrification, Solnit observes that it is 'homogenization, a loss of complexity, rather than absolute removal that most complained of' (Solnit and Schwartzenberg, 2002, p. 83). The place-image of an East London street market like Ridley Road, full of everyday shoppers from migrant and working-class communities, is increasingly absent from discursive presentations of East London. It is therefore tempting to read contemporary media, political and social discourses of discipline regarding the 'unruly space' of Ridley Road (Rhys-Taylor, 2014) as the mixophobic yearning of the gentrifier or the creative class for 'a territory free from that jumble and mess' of heterogeneity (Bauman, 2011, p. 63).

3   Markets and place-image

Rio de Janeiro

This chapter investigates the role of markets in narratives and images of place in globalized twenty-first-century cities. This is explored through two particular iterations of place – place-myth and place-image (Shields, 1991) – and in relation to markets in two of Rio de Janeiro's most visible sectors in terms of tourism, consumer culture, globalization and cultural heritage: the city's Zona Sul (South Zone) and the Projeto Corredor Cultural (Cultural Corridor Project) located in Centro. Rio de Janeiro is a city with an overriding investment in managing place-image because of its hosting of two global mega-events: the 2014 FIFA World Cup, and the 2016 Summer Olympic Games (Cummings, 2015; Jaguaribe and Salmon, 2012; Sánchez and Broudehoux, 2013). It is also, Beatriz Jaguaribe (2014) says, a city of landscapes, which, one might surmise, lends itself to being conceived as place-image. She writes, 'the prestige assigned to Rio de Janeiro's topography was built over centuries as the gaze towards the landscape was constructed according to a variety of interests, discourses and agendas' (2014, p. 2). How does that gaze perceive the city's markets, and how does that perception contribute to globally distributed and consumed place-images of Rio de Janeiro?

Markets in Zona Sul

Rio de Janeiro's Zona Sul covers the Atlantic coastal area of the city stretching south from Guanabara Bay to the Tijuca Massif, and incorporates neighbourhoods that belong to both informal and formal sectors of the city's built environment. Many of Rio de Janeiro's tourist attractions are concentrated here, including the iconic Christ the Redeemer statue, the photogenic peak of Pão de Açúcar and the globally recognized beaches of Copacabana and Ipanema. Zona Sul is therefore pivotal in the construction, communication and regulation of Rio de Janeiro's global place-myth as a laid-back, yet dynamic pleasure-ground for the expression and consumption of local *carioca* culture. Zona Sul is a visual metonym for the city that carries symbolic and real currency, and its place-image is subject to oversight and surveillance in discursive and material ways that place-images in other parts of the city might not be.

Formal and informal street-selling is well-integrated into the local consumer ecology in Zona Sul, even though the dominant groups of consumers in these areas belong to Brazil's most affluent socio-economic demographic groups,, Class A or B (or are employed as domestic workers to shop on their behalf), and are thus more likely to purchase goods from the significantly more expensive shopfronts and chain stores. Goods sold in supermarkets and chain stores – places of consumption frequented by the lower and middle classes in the US, Europe and Australia – are outside the financial wherewithal of many Brazilians. To be able to afford the prices of products in shops is a mark of social status and sacrifice (Oliven and Pinheiro-Machado, 2012), and inexpensive commodities of the sort bought at informal and formal street markets are associated with the consumer practices of the poor and the working classes.

There are, nevertheless, markets along many of the main shopping streets of Zona Sul, which operate out of temporary and permanent market-style installations and which offer everyday services and products, such as key-cutting, undergarments and snacks. There are also a number of weekly produce markets that move around the Zona Sul neighbourhoods on alternate days. These take over the public squares set back a block or two from the beach and are proximate to local shopping districts. The produce markets are well-frequented and are under the purview of local governance. In addition, there are *camelôs*, who are sole-vendors with a patch on the streets of the beach suburbs' residential and business districts, and *ambulantes*, a term used to refer to the itinerant vendors working the *orla* (waterfront). Both can be licensed or unlicensed. The difference is that *camelôs* have a more-or-less fixed, regular pitch on a particular sidewalk, whereas *ambulantes* are mobile.

The two beachside precincts, Copacabana and Ipanema, also host 'artisan' markets, which are administered by the Prefeitura da Cidade do Rio de Janeiro (the municipal council) and service the domestic and international tourist trade with souvenirs and handicrafts. Ipanema's Sunday 'Hippie Fair' (which was, as its name suggests, set up in the late 1960s as a counter-cultural meeting place) is a global tourist destination and was at one time listed as the top shopping attraction in the city on the TripAdvisor travel website (Hippie Fair Crafts Market, Rio de Janeiro n.d.). In Copacabana, a nightly market has set up along the *orla* since the 1980s on the traffic island in the middle of Avenida Atlantica. During the day, there is another tourist market with similar stock further up the beach at Praça Do Lido. The tourist markets and the weekly neighbourhood farmers' markets are authorised by the Prefeitura, and through their contribution to the formal urban economy are permitted to temporarily occupy valuable space in the street. Consequently, the spatial and material infrastructure for these markets is clearly demarcated and well-established.

Mobility and spatial practice in the Zona Sul is highly regulated through municipal and privatized operations of surveillance and security, yet informal markets are highly visible and, as in other parts of Rio de Janeiro, are a socially, if not always legally, tolerated part of the urban scene of consumption. They can exist as a string of operators servicing a contingent event such as a long bus queue or a sporting event crowd, or a temporal event like the evening rush at the elevator that services the Cantagalo community on the cliffs behind Ipanema. Alternatively, unlicensed *camelôs* cluster on the fringes of the formal markets, often replicating – and undercutting – the services and products sold in the regulated areas, or providing services that are complementary or auxiliary to those offered in the market. (One theory on the etymological origin of the word *camelô* is that it is borrowed from the French *camelot*, meaning 'merchant of low-quality goods'.) Along the fringes of both Copacabana tourist markets, unlicensed *ambulantes* spread out further and spill over onto the beach proper. Some sell *cangas* (sarongs), which they lay out on the beach, piling sand on the corners to stop them blowing away in the breeze. As a souvenir for day-trippers and holidaymakers, *cangas* are linked to the beach, materially and symbolically. They are both useful and sentimental. For vendors, they are comparatively easy to transport, exhibit and pack up – either at the end of the day, or more hastily when the authorities come around. They are the ideal, lightweight commodity for buyers and sellers alike.

Markets in the Projeto Corredor Cultural

Rio de Janeiro's Projeto Corredor Cultural (Cultural Corridor Project) is a central urban precinct, geographically delineated because of its heritage and cultural attributes, and protected by municipal legislation since 1984 (del Rio and de Alcantara, 2009; Brandão, 2006). Brandão, in talking about the successes and innovations of the Corredor, points out that the fostering and conservation of intangible qualities such as place and atmosphere were objectives of the project:

> The [Projeto Corredor Cultural], which has become a model of urban preservation in Brazil, recovered the symbolic meaning of this part of Rio de Janeiro, by giving back its cultural role that was built up over the centuries. A notable part of the city centre has, nowadays, an atmosphere that invites people to spend time there and it would not be an overstatement to say that central Rio de Janeiro has been one of the most attractive options for passing leisure time in the city for the last decade.
>
> (Brandão, 2006, pp. 45–46)

The liveliness and atmosphere of markets as a tangible cultural asset was explicitly taken into account in the project; in particular the market culture of Saara district, a tangle of streets where goods spill out of shops in a bazaar-like atmosphere – even though street selling is technically not permitted (da Cunha and de Mello, 2014). The name of this area is derived from the acronym SAARA (Sociedade de Amigos das Adjacências da Rua da Alfândega [Association of the Friends of the Rua da Alfândega Neighbourhood]), an association set up in the 1960s to represent stakeholders in the area. 'Saara' is also a homonym in Portuguese for 'Sahara', which is used linguistically as (geographically incorrect) shorthand for the area's social diversity and its concentrations of Lebanese, Syrian and, more recently, Chinese migrants (da Cunha and de Mello, 2014). The establishment of SAARA, and the eventual implementation of the Corredor, were interventions explicitly designed to alleviate the effects of major urban renewal projects in Centro, such as the removal of small-scale commercial and residential properties for the construction of major roads.

On the fringes of the Saara is a vast market facility selling cheap goods. The Mercado Popular Uruguaiana, also known as the Camelôdromo, was opened in 1994 as a formalized space of around 1,600 stalls for relocated *camelôs* who had previously worked the streets of the Corredor. Despite the attempt to regulate commerce and practices in the market, the informal economy is still prevalent. At the other end of the scale, another market in the west of the Corredor is active in gentrifying the precinct. On the first Saturday of the month, the Feira do Rio Antigo runs along Rua do Lavradio, a picturesque, semi-pedestrianized street in Lapa. The street is lined with colonial-era shops and residences, many of which have been converted into antique and vintage furniture shops. This vintage and craft market is not dissimilar to markets in other globalized cities that deploy the creative industries for urban regeneration and placemaking.

Another 'vintage' market with a very different atmosphere was also located in the cultural corridor. Every Saturday from 1979 until the end of 2013, the Feira de Antiguidades da Praça XV (Praça XV Antiques Fair) set up in a channel of land hemmed in on both sides and from above by roadways. Like second-hand markets across the globe, Feira da Praça XV utilized otherwise unused urban space. The flea market took its name from a nearby square, Praça XV de Novembro, which is a national monument and a tourist destination. The central location of the market was therefore noteworthy in that flea markets dealing in second-hand goods and junk are frequently pushed to urban peripheries because of their trade in the mucky, unpleasant business of waste.

Figure 3.1   Feira de Antiguidades da Praça XV, Rio de Janeiro

Feira da Praça XV's location and its history are tied up in the making and unmaking of urban space. In the early 1960s construction started on the Perimetral elevated road as a measure for Rio de Janeiro's increasing automobility. The road cut straight through the city's impressive Municipal Market, which was built in 1907 in the European iron and glass pavilion style, with four octagonal glass towers imported from Antwerp. The market was subsequently pulled down in stages. The demolition of the market produced a wasteland in the precinct by replacing human-scale pedestrian and commercial infrastructure with a roadway that cut off the port and harbour from the central business district. Over the next decades, the area's previous function as Rio de Janeiro's central market was metonymically remembered through one remaining corner tower (housing the Albamar restaurant, which continues to trade today), a square named for it on the site (Praça Mercado Municipal) and the emergence of two new markets in the area in the late 1970s: Feira da Troca, effectively a swap meet that functioned through barter, and Feira do Albamar, a more conventional showcase for antiques dealers. The establishment of the markets did not improve the amenity of the area, and the latter market moved to the well-heeled suburb of Gávea in the mid-1980s, leaving Feira da Troca to evolve into Feira da Praça XV (Freitas, 2013).

The market's terrain within the Projeto Corredor Cultural, and its later inclusion within the catchment area of the Porto Maravilha plan for the urban renewal of Rio de Janeiro's port, ensured that this cramped space would not be overlooked indefinitely. Rumours about the market's relocation dated from at least my first visit in 2008 (which was prior to the announcement that Rio de Janeiro would host the 2016 Summer Olympic Games), and were still circulating when I returned in May 2013, the year before the city hosted a number of 2014 FIFA World Cup games. The prevailing demands of urban infrastructure development driven by the global mega-events finally led to the demolition of the freeway to facilitate rehabilitation of the area. Despite the apparent inevitability of relocation, some stallholders only found out that the site was no longer available when they turned up one market day to find that the dismantling of the Perimetral had already commenced (Freitas, 2013).

The market has since been relocated into Praça XV proper and it remains to be seen if all stakeholders in the market, and in the city more widely, will benefit equally from the 'cleaning up' of the market and its surrounds. It has been renamed Praça do Mercado Feira de Antiguidades, signalling a separation from the previous market. It is spatially very different too, in that it spreads out in the square instead of being confined to a linear formation. Given its location in the city's historic square, it is subject to increased regulation in its operations and presentation. However, the increased exposure of this iteration of Feira da Praça XV is not about the city's transformed relationship with waste, rather it seems to occlude waste's presence even more by presenting second-hand goods as a statement of cultural capital or taste, rather than a material necessity or consequence of consumption.

Place-image and place-myth in Rio de Janeiro

In his well-known statement about stories, place and space, de Certeau refers to a type of making in the city that is linguistic and/or textual in its becoming: 'Stories… carry out a labour that constantly transforms places into spaces or spaces into places. They also organize the play of changing relationships between places and spaces' (de Certeau, 1984, p. 118). In a sense, de Certeau could be talking about the formation of what Rob Shields (1991) calls place-myth. Place-myth, which is made up of place-images, is a story about place that uses imagined and actualized images of that place.

> Images, being partial and often either exaggerated or understated, may be accurate or inaccurate. A set of core images forms a widely disseminated and commonly held set of images of a place or space. These form a relatively stable group of ideas in currency, reinforced by their communication value as conventions circulating in a discursive economy […] Collectively a set of place-images forms a place-myth.
>
> (Shields, 1991, p. 60)

Place myth and place-image are an iteration of place where 'Place… is both text and context' (Knox, 2005, p. 2). As they are theorized by Shields, they function somewhat like collective social imaginaries (Taylor, 2004) in that they have a legitimating function:

> [P]lace-images come about through over-simplification (i.e., reduction to one trait), stereotyping (amplification of one or more traits) and labelling (where a place is deemed to be of a certain nature). Places and spaces are hypostatized from the world of real space relations to a symbolic realm of cultural significations. Traces of these cultural place-images are also left behind in the litter of historical popular cultures: postcards, advertising images, song lyrics and in the settings of novels. These images connected with a place may even come to be held as signifiers of its essential character. Such a label further impacts on material activities and may be clung to despite changes in the 'real' nature of the site.
>
> (1991, p. 47)

There are a number of things I want to take from Shields' ideas. First, Shields is deploying 'image' as a 'generic term' (1991, p. 12), not as a solely visual phenomenon, which means that place-image can be produced and apprehended through speech, sound, touch, affect, memory and so on. Second, place-image in the first instance belongs to the representational order; it is a symbolic stand-in for a situated place. Place-image is therefore both indexical and metonymical. In this function, it can be mimetic or interpretive. However, with the growth of place as a marketing and branding device for cities (Anholt, 2010), devising and maintaining place-myth and place-image becomes more crucial, and there is commensurate expectation from the producers and consumers of that place-myth that places will conform to their place-images. In this feedback loop, place-image shapes place. Third, Shields does not clearly distinguish between place-images that are subjective or objective. Nevertheless, place-images are contingent upon and relational to who is producing, distributing and/or consuming the image. Lastly, place-image is open to transformation, as Shields elaborates:

> [T]here is a constancy and a shifting quality to this model of place- or space-myths as the core images change slowly over time, are displaced by radical changes in the nature of a place, and as various images simply lose their connotative power, becoming 'dead metaphors', while others are invented, disseminated and become accepted in common parlance.
>
> Opposed groups may succeed in generating antithetical place-myths (as opposed to just variations in place-images) reflecting different class experiences.
>
> (1991, pp. 60–61)

What this points to is that it is possible for multiple place-images produced by diverse actors to be temporally and spatially synchronized, although this does not necessarily mean that hierarchies of power are undone or resisted, as certain place-images and, consequently, place-myths will come to dominate based on their material and symbolic discursive power.

As Shields explained, place-myths and the place-images that contribute to them may be conflicting and competing. For example, one dominant place-myth that is reproduced globally and locally about Rio de Janeiro is of a glamorous urban playground for the adventurous, pleasure-seeking tourist. This place-myth is partially derived from the place-image of Rio de Janeiro's beaches as pluralistic, polyglot and accessible spaces that are representative of the inclusiveness of Brazilian society (Godfrey and Arguinzoni, 2012). Another place-myth of Rio de Janeiro is that it is an urban morass of informality, corruption, crime, and extreme social and spatial stratification. This latter place-myth has been contested through interventions such as 'favela chic', and the growth of poverty tourism in Rio de Janeiro's informal settlements (Cummings, 2015; Freire-Medeiros, 2013; Frenzel et al., 2012; Jaguaribe, 2014; Jaguaribe and Salmon, 2012). The enhanced visibility that accompanies the transformed place-image of a previously dangerous, poverty-stricken no-go zone is a visibility that prepares an area for gentrification (Zukin, 1995). The new place-image of barrios in Zona Sul as vibrant neighbourhoods ripe for gentrification is the result of ongoing violent, militarized state intervention since 2009. It was no spatial coincidence that the Zona Sul barrios, adjacent to the valuable visual and spatial economies of Copacabana and Ipanema, were among the first to be occupied by the UPP (Police Pacification Units).

Conflicting place-images might lead us to think of Rio de Janeiro as a heterotopology, which Foucault theorized '[a]s a sort of simultaneously mythic and real contestation of the space in which we live' (Foucault and Miskowiec, 1986, p. 24). Similarly, the place-images produced by and about the urban market are fraught with the contradictions of heterotopia, where synchronous and/or competing narratives of place emerge from continually reorganizing past, present and future of city, place and market in a 'meshwork of entangled lines of life, growth and movement' (Ingold, 2011, p. 63). For anthropologist Roberto da Matta (1991), the conflicted relationship to public spaces that Rio de Janeiro's official place-image acts out is surely a rehearsal of what he called 'the Brazilian dilemma'. According to da Matta, this is a series of binaries in the Brazilian collective imagination, one of which stigmatizes the street as the dangerous and unpredictable other to the home. The street is the sphere of 'punishment, struggle, and work,' and is conceived 'in Hobbesian terms: it is the way of all against all until some form of hierarchical principle can surface and establish some kind of order' (1991, p. 65). Not everyone perceives the street this way. For informal street vendors, the street can present possibilities for livelihood and community, for the creation of the 'subaltern urbanism' that Roy (2011) advocates.

Similarly, the place-images of markets can vary from context to context. Markets can be picturesque, so much so that they are used as devices for placemaking, place marketing, urban renewal and tourism. Yet they can also be confronting in that they materially manifest what urban management systems are often designed to move out of sight. Some markets bring to the fore matter, practices and communities that city officials would prefer to keep off-stage and unacknowledged. A flea market reminds us of the things we've thrown away, as Michelle de Kretser (2009) reflected on a visit to Melbourne's Camberwell flea market:

> Some people are immune to the second-hand aesthetic. To them, used goods aren't objects enriched with history, but signifiers of poverty and a lamentable lack of hygiene. Like history, shopping happens twice: the first time as fashion, the second as trash. When an object is new, its message of status or wealth, sophistication or cool is broadcast to all… The market offers a carnivalesque parody of shopping. It mocks our lust for acquisition, even as it plays to it. These acres bear witness to the scope of our greed and the fickleness of our desires. Whatever else a flea market might be, it's certainly a memento mori. And who wants a death's head in view when you shop?
>
> (de Kretser, 2009, p. 16)

For the reasons that de Kretser touches upon, markets – particularly informal and second-hand markets – have a fraught relationship with visibility. To have markets that remind us of waste, poverty, or even death, visible and freely operating in spaces for consumption can be interpreted as a failure of urban governance (Hiebert et al., 2014). Increasingly, central wholesale and everyday markets have been moved to peripheral spaces in cities because their infrastructure might be considered visually unsuited or aesthetically unpleasant to the new or transformed uses to which an urban centre is being put (see Jacobs, 1996). One cited rationale for relocation from central sites is that these markets are unsightly and messy. They have joined second-hand and flea markets that traditionally occupy urban edgelands because of their trade in waste and their association with informality (Mörtenböck and Mooshammer, 2008).

Informality and place-image

Informal and/or secondary circuits of commodity exchange, particularly those that manage waste, are indispensable for urban livelihoods in Brazil and elsewhere across the globe (Coletto, 2010; Evers and Seale, 2014) and for socially and materially sustainable cities (United Nations Human Settlements Programme, 2010). While it may be the case that there has been a 'global transformation from a "modern" economic/political system, which saw street vendors and the informal sector as parasitic or at best inefficient, to a "post-modern" economic/political system in which street commerce is often seen as a source of growth and flexibility' (Cross and Morales, 2007, p. 2), structural inequalities between informal and formal have not been addressed. Legal, political and media discourses that stigmatize and in many cases criminalize street commerce in order to maintain place-image pose serious challenges for those trying to make a livelihood in markets. It not only affects individuals. Treating markets that are informal or that deal with waste as abject or illegal directly inhibits the design and development of socially and environmentally sustainable and resilient cities.

Waste and informality are interconnected in that they both belong to shadow and interstitial spaces and economies in the city. Their vulnerability to the currents of local and global governance and development is compounded by the structural inequalities of a hierarchical informal/formal binarism that deprives those in the informal sector of economic and political power. Since Keith Hart's (1973) seminal, dualist study of informal employment in Ghana, which coined the phrase 'informal economy', Diego Coletto observes that, by and large, 'definitions of the informal economy are constructed in opposition to the formal or regular economy, so that the informal economy is either an intermediate stage, an obstacle, an instrumental appendage, or an alternative to the dominant model of economic development' (2010, p. 16). At times it is conflated with criminal activity.

Casting the informal as the other to the formal implies that the two are in binary arrangement. However, studies of informal economies have shown the boundaries to be discursive. They do not hold up to scrutiny when one examines actualized circulations of people and goods through the imagined borders of informal and formal (Coletto, 2010; Evers and Seale, 2014). Moreover, for many who work in informal economies, discursively constructed disrepute does not align with their own perception of their work and community, nor does it take into account the vast numbers of urban dwellers dependent on the wide range of employment opportunities available within the informal sector (Abdulazeez and Pathmanathan, 2014; Neuwirth, 2012; Seale and Evers, 2014). 'Informal' connotes something that is not formed, or that lacks shape, whereas informal economies are frequently structured according to a neo-liberal capitalist model of laissez-faire free-market individualism, entrepreneurship and self-regulation, with responsibilities and costs of employment devolved to the individual (Venkatesh, 2006, 2008). Workers in the informal sector are therefore a core set of workers in the global economy. They are hardly marginal in number, even though they might be marginalized through lack of political or symbolic representation in the urban economy. Furthermore, their growing numbers are promulgated by globalized neo-liberalism, in that 'less formalized economic activity operates as part of the coping strategies of people in communities caught by... changes in the global economy' (De Bruin and Dupuis, 2000, p. 53; see also Breman, 2013). Workers in the informal economy who may previously have been considered the outcasts of modernity (Bauman, 2004) are now the embodiment of a neo-liberal politics that coalesces around the primary attributes of contingency, mobility and flexibility in global labour.

Formal urban economies in Rio de Janeiro, such as the network of official markets in Copacabana and Ipanema, regulate flows of bodies, labour and capital. By contrast, the movement and stasis of informal markets and vendors is *responsive* to flows of people, goods and communication in urban space (Hepworth, 2014). This responsiveness to flows is surely indicative of subaltern urbanism, which Ananya Roy says is 'an important paradigm, for it seeks to confer recognition on spaces of poverty and forms of popular agency that often remain invisible and neglected' (2011, p. 224). As economically, socially and spatially marginalized as *ambulantes* and *camelôs* may be in Rio de Janeiro, they are deploying the increasingly restricted possibilities of urban public space for a livelihood.

On Ipanema and Copacabana beaches, amongst the many *cariocas* and tourists walking, cycling, sitting with an *agua de coco*, playing beach volleyball or *frescobol*, are working people at the *barracas* and the *quiosques*. There are also licensed and unlicensed *ambulantes* who sell favourite local snacks and beverages, such as *mate*, *açai*, Globo crisps, sandwiches and peanuts as well as global brands of soft drink and ice cream. Some itinerant vendors sell site-specific products – swimming costumes, resort wear, sun hats, and sun cream – while others sell cigarettes, temporary tattoos and silver and beaded jewellery. The emphasis on glamorous consumption in the visual and material economies of the beaches brings into relief the division between those who work on the beach and those who do not. The presence of some sellers, mostly informal, who are visible reminders of 'advanced marginality' (Wacquant, 2008) causes a rupture in the smoothness of any unilateral meaning of the beach as an egalitarian space of/for consumption, and reinscribes the tensions 'between social order, status, and hierarchy on one hand, and democratic rights, social diversity, and accessibility on the other' (Godfrey and Arguinzoni, 2012, p. 18).

Given the prevalence of informal markets in Rio de Janeiro, and consumers' apparent ease with their presence (played out through phatic and commercial exchange), it could be surmised that informal markets and vendors contribute to the place-image of the city. Informal street selling is clearly a grounded, everyday practice; unexceptional in spite of governmental policies and operations that attempt to exceptionalize it. Even so, this place-image is contested and countered by the signals and instruments of another place-image along the *orla*. In 2010, tents appeared on the beach to support and promote Choque de Ordem (Shock of Order), a militarily enhanced strategy for bringing 'pacification' to favela neighbourhoods. The citywide Choque de Ordem stretched to the Zona Sul beaches in the form of a crackdown on unlicensed vending. The tents on the sand at Copacabana and Ipanema were also an instigation to moral panic surrounding the use and practice of public space in Rio de Janeiro. Through the control of flows of people and goods along the beach, or at least giving an impression that control of flows had been achieved, the Prefeitura's objective is a place-image of a beach secure from informal activity (which in this case is discursively conflated with criminality).

This strategy is supported by the built environment along the beach. Copacabana is one of Rio de Janeiro's four designated 'Olympic Zones' at the 2016 Summer Games. Newly constructed kiosks on the beachfront have replaced the more shack-like bars and their ramshackle plastic chairs and red and yellow umbrellas. The new glass and steel structures are more permanent and 'café-like', with semi-cordoned-off, undercover eating areas that clearly define privatized areas of consumption. The barriers work socially, as well as spatially, protecting the clientele from the *ambulantes* and panhandlers who roam the beach precinct.

Waste and place-image

As mentioned previously, Rio de Janeiro's flea market has undergone a 'clean-up' as part of urban renewal projects connected to planning and infrastructure for mega-events. The previous market's conspicuous display of waste in the street resisted hegemonic projections of what constitutes liveability in urban contexts (Coletto, 2010). Rio de Janeiro's flea market was 'matter-out-of-place' (Douglas, 1966) and at odds with the official place-image of the historic, touristic Cultural Corridor, and of Rio de Janeiro itself as a modernized global city. The market was a visible congregation in the city's political, financial and cultural centre of 'urban outcasts', communities who are usually pushed to the social and spatial peripheries of cities (Wacquant, 2008).

Our notion of what is waste varies from society to society, is dependent upon our material circumstances, and differs according to our position within a social order. Attitudes towards waste can mark cultural variance or similitude. Proximity to or distance from waste is often based on class. What ideas and practices to do with waste frequently share is an objective to sanction, segregate, marginalize and discriminate. John Scanlan observes, 'the meaning of waste carries force because of the way in which it... operates within a more or less moral economy' (2005, p. 22). Indeed, waste management as a discourse where the unwanted is firstly separated and contained, and then of use once more is so symbolically powerful that it is metaphorically deployed in a number of social and political contexts (Bauman, 2004); for instance 'work-for-the-dole' programmes whose ideological function is to recoup labour capital from 'wasteful' welfare expenditure.

In *Purity and danger*, Mary Douglas observes that the desire to install order leads to 'ideas about separating, purifying, demarcating and punishing transgressions [that] have as their main function to impose system on an inherently untidy experience. It is only by exaggerating the difference between within and without, above and below, male and female, with and against, that a semblance of order is created' (1966, p. 4). It is the ruptures or blockages in management systems that produce 'matter-out-of-place'. The unease that occurs when waste confronts us with its materiality emerges from expectations regarding order. We are therefore socially and culturally pre-disposed to view waste pejoratively (Elias, 1978; Laporte, 2000). Some of our rationale for marginalizing it may have sound physiological basis. Nevertheless, waste is an obligatory, insistent and, above all, valorized component of globalized, neo-liberal capitalism. Waste is neither abject, nor excessive; rather it sustains capitalism's growth. We might even say, as David Trotter does, that in capitalism 'the success of the enterprise can be measured by the waste-matter it produces, by the efficiency with which it separates out and excludes whatever it does not require for its own immediate purposes' (2000, p. 22). The disconnect between waste's symbolic role and waste's actualized role in global capitalism is the paradox of waste.

The precarious conditions and situations of those who work with waste, and the remote urban locations to which they are often relegated are based upon symbolic representations of waste as marginal and abject, and do not reflect waste's actual value within local and globalized economies. Instead, informal and formal second-hand markets are nodes within a globally ubiquitous network of secondary economies that generates valuable social, economic and material infrastructure in cities (Evers and Seale, 2014; United Nations Human Settlements Programme, 2010). At the end of each day, very little rubbish remains on the beaches of Rio de Janeiro, or on the streets of its affluent and middle-class suburbs. Through the night and early morning, phalanxes of sanitation workers and scavengers, working in both the informal and formal economies, sort and clean up much of it.

Some of that rubbish is handpicked and reclassified as waste, and bound for secondary markets, like Praça da XV, where it can be sold and bought anew

(Coletto, 2010). The vendors at Praça XV are entrepreneurial, reincorporating waste back into circuits of exchange in a process that provides employment and waste management for the city. This is the second-hand market's value in spite of its discursive positioning within the representational and material orders of the city. Conceptually, empirically, waste and commodity are two sides of the same coin – as the trade at Feira da Praça XV illustrates. The distinction between waste and commodity is not material; rather, their classification as one or the other hinges on how we treat or deal with them. Feira da Praça XV mobilizes this phenomenological overlap. The waste that is sold there is a commodity.

Figure 3.2   Inventory, exhibition, commodification

Feira da Praça XV, as it moved through formal and informal articulations, was always subject to numerous practices of ordering carried out at multiple micro- and macro-levels of the market. The infrastructure at the end closest to Praça XV was that of a licensed market. Vendors with more expensive stock in better condition had uniform stalls provided by central management. The regulated parts of the market at this end were more spread out, and consequently, more pleasant, spatially and aesthetically. Sometimes their collections were curated and carefully arranged. This was the closest the market came to the phenomenon of fetishized vintage (Gregson and Crewe, 2003; Palmer and Clark, 2005). Some were highly specialized in their collections. One stall sold only vintage surplus pencils in their original tins and boxes; another stall was heaped high with (to my mind) sinister-looking used steel surgical instruments.

As you moved away from this section of the market, the space under the Perimetral narrowed, and so, too, the space between aisles, stalls and people was reduced. Here there was a second tier of stalls with red-and-white striped canvas shades instead of green. The wares on display in this section were more eclectic, thrown together, visually messy. Things might be broken, missing a part, or be the single remnant of a pair. The flyover's pylons functioned as de facto markers signalling increasingly ad hoc spatial organization and goods until the market trickled out with groups of sellers who displayed their goods on a towel or directly on the ground. On the fringes of these final sections were groups of Rio de Janeiro's globally renowned *catadores* (see Lucy Walker's 2010 documentary *Wasteland*) who sorted through the leftovers. Even here, where the market seemed more like a scrapyard than the stage for consumption, there were processes of sorting and ordering at work.

Feira da Praça XV instigates order in an arena where many assume there is none to be found. It reinstitutes order among previously discarded objects through inventory, exhibition and, above all, commodification. In Rio de Janeiro, place-images and place-myths produced and circulated to attract global flows of capital, including those connected to mega-events and global tourism, are based on the same logic of inventory, exhibition and commodification at work at the Praça XV market. However, it is unlikely that we will be seeing representations of informal markets or flea markets that deal in waste in this set of place-images. In spite of this, these markets are very much part of another set of place-images of Rio de Janeiro that emerge from everyday making. In the next chapter, I want to think a bit more about markets and place within the context of the everyday, and in particular the constraints on markets as everyday places in globalized cities.

4 Markets as unremarkable and remarkable places

Paris, Amsterdam, Hong Kong, Beijing

For large numbers of the global population, markets are unremarkable places. At the same time, some markets are being made or remade as remarkable places through interconnected discourses of tourism, place marketing, cultural heritage, and the Creative City. The (re)invention of markets as remarkable places is one response to globalization, which Paul Knox says, has 'prompted communities in many parts of the world to become much more conscious of the ways in which they are perceived by tourists, businesses, media firms and consumers. As a result, places are increasingly being reinterpreted, reimagined, designed, packaged, themed and marketed' (2005, p. 4).

In this chapter, I explore the tensions between remarkable and unremarkable place that occur when markets are appropriated by globalizing discourses. To do this, I want to look at four globalized cities: Paris, Amsterdam, Hong Kong and Beijing. All four are well-established on global tourist itineraries. According to one ranking system, Hong Kong is ranked number one globally in terms of numbers of international tourist arrivals, Paris is number five, and Amsterdam and Beijing are 27 and 34 respectively (Top 100 City Destinations Ranking, 2015). The prevailing place-myths and place-images of these cities are therefore familiar to a global audience, and have an impact on flows of global tourism and global cultural consumption.

Un/remarkable places

In Ridley Road Market in East London (about which I wrote in Chapter 2), Alex Rhys-Taylor detects a dialectic at work between remarkable and unremarkable. This street market, Rhys-Taylor writes, is 'the site of some of the most remarkable cultural activity undertaken by the city's inhabitants. It is a space through which transnational connections are made, and out of which new local cultural formations emerge. As remarkable as it is, however, much of that activity appears at first to be banal' (2014, p. 45). The ordinary in Ridley Road is produced from commonplace diversity in a super-diverse city (Wessendorf, 2014). In this urban environment, as in many other cities, cultural difference is hardly remarkable. It is part and parcel of the everyday, and to situate it otherwise risks turning it into a staged scene of cultural diversity (Bhabha, 1994; Gilroy, 2004). However, Ridley Road is remarkable precisely because making place in this market is ordinary. When I say that this is remarkable I do not mean that ordinary forms of making or the production of ordinary place in markets is remarkable. What I mean is that given the pressures on markets and place in globalized cities, Ridley Road's achievement is that is has remained a local place produced by making that is quotidian and ritualistic. This market has, so far, resisted being remade as a local place for global consumption, as has happened in nearby Broadway Market.

As an ordinary place, Ridley Road Market 'provides an arena in which everyday, "common-sense" knowledge and experience is gathered; provides a site for processes of socialization and social reproduction; and provides an arena for contesting social norms' (Knox, 2005, p. 2). Knox is getting at something else in his description of ordinary places: their collectivity. For Ben Highmore,

> this sense of collectivity is central to thinking about the ordinary. While the everyday might be an endless succession of singularities it is not helpful to understand it as peopled by monads. The ordinary harbours an abundance that is distinct from material plenty: it is there when we talk about something as common, it is there when we talk about society, and it is there when we talk about 'us'.
>
> (2011, p. 5)

The everyday market, as an everyday place, is imagined and practiced collectively. Jacques Augoyard noticed this in his ethnography of a newly built urban housing development in Grenoble. Augoyard observed that when the town centre was retrofitted with a market, a collective sense of appropriation of the market as a place was affirmed linguistically through the widespread, shared adoption of an informally occurring name:

> Although it refers to a country-village-like feature that contradicts the urban setting of this neighbourhood, the term *Market Place* nevertheless was established spontaneously and appears, at the time of the market's creation, in the narratives of all inhabitants, irrespective of social group, age, or sex. The redeployment of an ever-so-common term has therefore marked the site according to its temporary function and with an indistinct appropriation performed on the scale of the entire neighbourhood, as by general consensus.
>
> (Augoyard, 2007, p. 87)

Augoyard's example supports Knox's statement about a dynamic, adaptive relationship between people and ordinary places:

> As people live and work in places, they gradually impose themselves on their environment, modifying and adjusting it to suit their needs and express their values… People are constantly modifying and reshaping places, and places are constantly coping with change and influencing their inhabitants.
>
> (Knox, 2005, p. 3)

This adaptability of place was suggested by Shields' (1991) theory of place-image in the previous chapter as well.

However, Knox is cautious about place's mutability in a 'world of restless landscapes in which the more places change the more they seem to look alike, the less they are able to retain a distinctive sense of place' (2005, p. 3). This line of thought about placelessness as a consequence of globalization is picked up by Val Plumwood:

> There is a serious problem of integrity for the leading concepts of much contemporary place discourse, especially the concept of *heimat* or dwelling in 'one's place' or 'homeplace', the place of belonging. The very concept of a singular homeplace or 'our place' is problematised by the dissociation and dematerialization that permeate the global economy and culture.
>
> (Plumwood, 2008, p. 139)

Plumwood is arguing for an ethics of place, one that is cognisant of the 'shadow places that provide our material and ecological support, most of which, in a global market, are likely to elude our knowledge and responsibility' (2008, p. 139), but her argument, and others like it, are problematic in that they unintentionally confirm the discursive binary between place and space that is perpetuated by the very capitalist thought and practice they seek to criticize. Here place is the rooted, intransigent, prickly, impractical, obstructive other to unbounded, free, constructive, pragmatic neo-liberal space. Linda McDowell (1997) has observed that place can be regulated by exclusions, which undoubtedly leads to what Doreen Massey calls 'problematical senses of place, from reactionary nationalisms, to competitive localisms, to introverted obsessions with "heritage"' (1994, p. 151). Due to its association with the customary, notions of place can veer towards the conservative and the nostalgic. Certainly when place is under contestation or under threat, the practices and ideas that mark place are territorial.

Yet if we return to Ridley Road Market, place is not produced unilaterally by one group. Place that emerges from making in Ridley Road is collective. The localism in Ridley Road is constituted by global flows, and is therefore hybrid and heterogeneous. As an urban market, Ridley Road's ambivalence between local and global is unexceptional because, as Stallybrass and White explain, markets exist at neither end of the local/global axis:

> At once a bounded enclosure and a site of open commerce, [the market] is both the imagined centre of an urban community and its structural interconnection with the network of goods, commodities, markets, sites of commerce and places of production which sustain it. A marketplace is the epitome of local identity (often indeed it is what defined a place as more significant than surrounding communities) and the unsettling of that identity by the trade and traffic of goods from elsewhere.
>
> (Stallybrass and White, 1986, p. 27)

It is, paradoxically, globalizing discourses such as cosmopolitanism, tourism and urbanism that situate markets as the most parochial of urban spaces. In the popular imagination, few urban sites are more resonant of local place. Travel guides and shows for global audiences frequently recommend markets as an 'authentic' experience of place. Defined by this schema, Ridley Road Market is an authentic scene of everyday diversity (Zukin et al., 2015) in London. However, authenticity is not emergent but discursive, and ultimately elusive because the act of seeking it out or bringing it into being bears with it the ironic seeds of [authenticity's] own destruction' (Coleman and Crang, 2002, p. 3). So-called authentic places are altered or even destroyed by globalizing processes such as tourism or place marketing at the very moment that they are being interpellated by them. This interpellation produces a remarkable iteration of place, which still lays claim to an authenticity that is based on that place, but too often crosses over into the themed environment (Miles, 2010). As Plumwood has written, 'if commodity culture engenders a false consciousness of place, this meaning can be fake' (2008, p.139). This 'false consciousness of place' may be one way to think about urban markets that are engineered to be remarkable through the operations of placemaking, those choreographed experiences that have specific and strategic uses and objectives with regard to place. As advanced in Chapter 1, atmosphere is emergent, not a product (Pink and Mackley, 2014), and attempts to conjure place and atmosphere in markets can end up colourless and flat.

Markets and proximity in Paris

Susan Parham calls markets 'the outdoor room' (2015, p. 71). Parham is using this image to reference the human-scale of markets (which I look at in more detail in Chapter 7) and their conviviality, rather than suggesting that the place of the market is somewhere that people feel at home in public space or, at the very least, familiar and at ease. Thinking about Parham's image in relation to markets in Paris, I am reminded of Walter Benjamin's observation in *The arcades project* that Parisians 'inhabited' the street as if it was their home and in doing so, 'Parisians make the street an interior' (Benjamin and Tiedemann, 1999, p. 421). Benjamin's remarks and Parham's metaphor are useful for thinking about proximity and markets in Paris. In spite of the design of their city being shaped by a profound distrust of the potential of public space (Harvey, 2005), Parisians nevertheless experience a proximity to the street, to public space, and to public places like markets that engenders familiarity.

Look down as you stroll along certain boulevards and streets of Paris' *arrondissements* and you will notice rows of metal slots punctuating the pavement. At least once a week, sometimes more, poles are fitted into these holes and slatted canvas awnings unrolled on top of them as the fixtures of local neighbourhood markets are installed. At these markets, which are regular events in

Figure 4.1 Rue des Pyrenées, Paris

every Paris neighbourhood, food makes up the majority of the goods on sale. The location of the markets in local streets provides proximity to consumer and food infrastructure. Paris' residents expect proximity to good quality, fresh produce, even though they are not immediately proximate to agricultural land. On Sunday morning the market in the Rue des Pyrenées in the 20e is filled with people, young and old. It is a produce market, although there is the odd stall selling quotidian objects such as kitchenware or socks. The street is closed off to traffic so walking assumes primacy (Shortell and

Brown, 2014). People can navigate the architecture of the market in an aleatory manner. You bump into and step around and pick up things.

Markets are places where proximity cannot be avoided – with the goods,
with other people, with the street, with waste. Proximity, as a material condition, is a characteristic of the urban experience of Paris in general: proximity
to public transport, to educational facilities, to cultural infrastructure, and to
local, small-scale, highly specialized food production. It can be surmised that
many of the shoppers and visitors are local residents or workers as the Rue
des Pyrenées market does not do anything remarkable as compared to the
other weekly markets in Paris. Those from further afield would have no need
to make a special visit to this market, unless there was a particular vendor for
whom they had a preference, or if they were unable to visit the market in their
own neighbourhood on the day it was held. The atmosphere is workaday, but
leisurely nevertheless. A pleasant place to do your weekly shop, as opposed
to a social occasion. One might even say that the spatial, social and material proximities that make place in Rue des Pyrenées are a demonstration of
an urbanism that is everyday, slow and sustainable (Farr, 2008; Haas, 2012;
Knox, 2005).

Markets are unexceptional in Paris whether it is the polite charm of Rue
des Pyrenées, or a noisy convocation in a less gentrified neighbourhood like
La Goutte d'Or in the 18e (Kaplan and Recoquillon, 2014, 2015), or the touristic *bouquinistes* (booksellers) along the Seine. Perhaps it is proximity and
complacency that permitted the renowned and architecturally influential Les
Halles central market – where 'the history of a building and that of a city
coincide' (Mead, 2012, p. 144) – to be torn down and replaced with a concrete
underground shopping centre in what may be one of the most derided urban
renewal projects of the twentieth century (Chevalier, 1994; Hazan, 2010;
Merrifield, 2014). In a city of markets, it is ironic then that of Paris' two most
famous markets, Les Halles is an archival memory, and the other, the Marché
aux Puces de Porte de Saint-Ouen (the original 'Flea Market' that gave its
name to all others) is not in Paris. The Marché aux Puces is situated on
the *banlieue*-side of the Péripherique, the city's clogged ring road – which
means that it is technically outside Paris. Paris' automobilized city limit
runs a circuit that roughly traces the gates and ramparts of its medieval
walls. Historically, *brocantes* or second-hand markets were located outside the city's gates (like the Porte de Saint-Ouen). This was a consequence
of the 'embourgeoisement' of the urban centre in the nineteenth century,
when 'insalubrious' industries like rag-picking were shifted outside the
city's perimeter (Harvey, 2005; Ratcliffe, 1992). The rag-picker's geographically and socially peripheral existence is visualized in Georges Lacombe's
1928 docudrama *La zone: au pays des chiffonniers (The zone: in the land
of the rag-pickers)*. Initially, Lacombe documents the rag-pickers' activity
within a landscape defined by tropes of the modern metropolis – newspa-
pers, crowds, the Paris Metro, streetcars. The *chiffonniers* empty out bins of
rubbish onto the city's streets and rummage through them. The group load

up their carts and move their salvage to the 'zone' of the title, a band of interstitial ground at the edge of the city. It is in this zone that flea markets like the Marché aux Puces sprung up.

It is unlikely that the rag-pickers of Lacombe's film would recognize the contemporary flea market standing in its place. Like Beijing's Panjiayuan flea market, which I look at later in this chapter, the Marché aux Puces has been sanitized. In fact, the market has moved beyond mere sanitization and is approaching peak gentrification. A restaurant designed by Philippe Starck has opened up, and a new boutique hotel from the owners of the hip Mama Shelter hotel in Belleville is slated to open in 2016. While Panjiayuan has been turned into a standardized consumer experience with the organization and regulation of a shopping centre, at Porte de Saint-Ouen there is just enough eccentricity and atmosphere to fuel the libidinal circuits of tourism in Paris. It is the first and only urban precinct to be listed by the French government as cultural heritage on the basis of its *ambiance* (atmosphere) (Félix, 2002). The Marché aux Puces is more a curated *Wunderkammer* (Siradeau, 2008) than junk shop, however. Today, searching the flea market is less *chiffonnier*-like rummaging (Trotter, 2000), not knowing what you might turn up and what its value might be, and more browsing an array of predictable finds whose value has already been predetermined.

The Marché aux Puces is, according to its website, the fourth most popular tourist site in the greater city, attracting five million visitors a year (4e site touristique de France, n.d.). It is, of course, also a site of consumption, and as such fits within the paradigm of shopping as leisure activity for tourists (Jansen-Verbeke, 1991). In 2014, according to figures from the city's Office of Tourism and Conventions, Paris had close to 22.5 million international and domestic visitors for touristic purposes (L'office du tourisme et des congrès de Paris, 2015, p. 9). In a sample survey of those tourists, close to 20 per cent reported visiting flea markets and markets during their stay (2015, p. 12) and almost 44 percent said they went shopping as a tourist activity. Some 14.6 percent of tourists who responded to a question about their reasons for visiting Paris cited shopping (2015, p. 12).

Shopping as a place marketing strategy for Paris (Rabbiosi, 2015) is more likely to promote French antiques stores, designer flagships along the Rue Saint-Honoré, small idiosyncratic boutiques, Belle Époque department stores, and remarkable markets in tourist precincts such as Rue Mouffetard in the Latin Quarter, the bird and flower markets on the Ile de la Cité, and of course the Marché aux Puces. Unremarkable markets in Paris do not feature prominently in globally distributed place-images of the city because they are not consistent with Paris' global brand as a luxury or cultural shopping experience (Rabbiosi, 2015). Instead, they are a destination for cheap, mass-market goods. An example is the Rue Dejean market in the 18e arrondissement, which

provides numerous opportunities for deeply discounted shopping. These are hardly the fancy boutiques, jewellery stores, and fine-food establishments that tourists associate with Paris... There vendors à la sauvette sell all manner of cut-rate items: things to eat like chestnuts and grilled corn, vegetables like eggplants, women's hand- bags, and pirated DVDs. "Stands" are constructed from old shopping carts, overturned cardboard boxes, crates, sacks on the street, and card tables.

(Kaplan and Recoquillon, 2014, p. 41)

Kaplan and Recoquillon report that the atmosphere in the Rue Dejean market and its surrounds is unexceptional because 'these are not organized festivals, but simply the actions of people coming together' (2014, p. 41) in a super-diverse city. Perhaps Rue Dejean's most remarkable feature, which it has in common with Ridley Road Market, is that it is an unremarkable place in a city that is being reconstructed as a 'Disneyland for the cultivated' (Merrifield, 2014, p. 28). In order to substantiate the authenticity of its remarkable place-image, Paris has become the city that Henri Lefebvre discovered in contemporary textual mediations of the urban:

The text is moving away. It takes the form of a document, or an exhibition, or a museum. The city historically constructed is no longer lived and is no longer understood practically. It is only an object of cultural consumption for tourists, for an estheticism [*sic*], avid for spectacles and the picturesque.

(Lefebvre, 1996, p. 148)

An unremarkable market in Amsterdam

Like Rue Dejean, Ten Katemarkt is an unremarkable market. It is located in Ten Katestraat in the Amsterdam neighbourhood of Kinkerbuurt. Every day, except Sunday, the market is installed and then dismantled by a team of workers in a finely tuned and intricate correspondence of bodies, skills and materials (Ingold, 2011). Ten Katemarkt has a standard mix of stalls selling inexpensive electronics and accessories, clothing, fresh produce, a couple of local favourite foods (Dutch cheese, frites), and 'ethnic' specialities (olives, spices). I went for a stroll with Aslı Duru, a colleague who researches markets in Istanbul, and we agreed that Ten Katemarkt is what we would classify as a successful market, in that it indicated 'dailyness and groundedness – what might be called the positivities of place' (Gibson-Graham, 2006, p. xxxiii). It is busy with people buying and chatting. It offers a decent range of reasonably priced goods that local residents would use in their day-to-day routines. To us as visitors, the market was a good example of urban intersubjectivity, which Knox locates as both everyday and collective:

The basis of both individual lifeworlds and the collective structure of feeling is intersubjectivity: shared meanings that are derived from the lived experience of everyday practice... Successful urban places – from

Figure 4.2   Ten Katemarkt, Amsterdam

both the insiders' perspective and an outsider's perspective – not only
have the lineaments of good urban form but also an underlying dynamic
of activity – routine encounters and shared experiences that make for
intersubjectivity.

(Knox, 2005, p. 2)

Ten Katemarkt is also a good example of what Sharon Zukin (2012) calls 'vernacular spaces' in her ethnography of another Amsterdam shopping street.
Unlike the ordinary shopping street that Zukin documented, Ten Katestraat,
in spite of being unremarkable, is not what the Dutch call '*[G]ezellig*, or
cosy, a term of approval that signifies the observer feels comfortable... *gezel-
lig* invokes an archetypal folk society of rural villages, a social life that exists
outside of – and in contrast to – the worldly city' (2012, p. 286). In fact,
Ten Katestraat and its market are a reminder of the heterogeneous world-
liness of cities. Street and market are shaped by local migrant communities
(Otterloo, 2009), and both are visibly diverse in their clientele, vendors and
the goods on sale. The shops that line the street service local Turkish and
Surinamese communities: New Chicken is halal; the Tropic 2000 store and
the Asian Caribbean Market both display Surinamese flags; the Toko Surima
eatery on the corner advertises that it is '*voor alles wat exotisch is*' ('for everything exotic'). Despite Toko Surima's reference to the 'exotic', the street is an

example of super-diversity that is embedded in the everyday. Unlike demarcated urban spaces, such as Chinatown precincts in some globalized cities, (Hall and Rath, 2007), Ten Katestraat and Ten Katemarkt do not deploy cultural diversity (Bhabha, 1994) as a placemaking or tourism strategy. Instead, they are intangible cultural heritage of the sort that Zukin (2012) says is often overlooked by virtue of its banality or lack of remarkableness.

The neighbourhood where Ten Katemarkt is located is also unremarkable in a globalized city. In an ostensibly textbook case of gentrification, Kinkerbuurt, which has in the past been a predominately working-class and migrant area, has experienced an influx of young people and professionals who seek lower rents and property prices close to the centre of the city (Lindner and Meissner, 2015). How long Ten Katemarkt can survive as an ordinary market in such an environment remains to be seen. The gentrification process in Kinkerbuurt has recently received a huge boost with an urban renewal project literally on Ten Katemarkt's doorstep. The recently opened De Hallen Amsterdam describes itself as a 'Centre for media, culture, fashion, food and crafts'. De Hallen Amsterdam is a product of, and in turn produces, a diffuse Creative City strategy that is being used to release pressure on the city's cramped central tourist and cultural districts. In 2015 it was reported that central Amsterdam's tourist areas, which are spatially limited due to the arrangement of the built environment, were reaching capacity in terms of visitor numbers, and the city was adopting a policy of building tourist infrastructure in neighbourhoods further afield (Reimerink, 2015). In support of this policy, the De Hallen complex includes an upmarket hotel, along with a media studio, cinema, a library, boutiques, workshops and showrooms. It is an attractive-looking series of making, eating and shopping spaces that fully utilize the post-industrial aesthetic and scale of the converted turn-of-the-century tram shed in which it has been built. In spite of its aesthetic appeal, the atmosphere was incomplete, perhaps because it hasn't quite become a place yet. Visitors appeared unsure about where they were going or how they were meant to be using and experiencing the space.

Of most direct impact to Ten Katemarkt are Food Hallen, De Hallen's food hall, and the regular market days held at the complex. Food Hallen is a market-like space where diners buy gentrified street food – *banh mi*, falafel, and burgers – for higher prices than if bought in an actual market. Even though the booths emulate market stalls, it is not a market, because the infrastructure and ethos is more akin to that of bricks-and-mortar establishments. Food Hallen is representative of a trend for 'restaurants, street merchants, and even supermarket chains [taking] inspiration from public markets, which therefore become a reference point for economic agents that do not share public markets' very ideology' (Visconti et al., 2014, p. 350). Food Hallen is probably closer to a hawker centre in Singapore in its bureaucratization of the market, but unlike hawker centres, the businesses are not former street vendors who have been warehoused in a 'modernizing' drive. Judging from the crowds there in the evenings and on weekends, Food Hallen is very popular.

Figure 4.3 De Hallen Amsterdam

Not so much with locals, however. My host in Amsterdam (an artist who had lived in Kinkerbuurt for eight years) complained that it was filled with people from elsewhere, and people who lived locally no longer found it 'cool'. There appeared to be little demographic overlap between Ten Katemarkt and the food hall, and there was more foot traffic in and out of the eastern Bellamyplein entrance than the entry point off Ten Katestraat.

De Hallen also hosts its own market every second weekend, the craft-oriented 'Local Goods Weekend Market'. Again, there was little cross-over with Ten Katemarkt outside in terms of customers or goods on sale, aside from some artisanal olives that were the upscale version of the olives being sold out on the street. One hand-printed children's T-shirt I looked at had a price-tag of €60. For now, De Hallen and Ten Katemarkt, and their respective clientele, are occupying separate spheres despite their spatial propinquity. What will be of interest is whether De Hallen brings new consumers to Ten Katemarkt, thereby 'reviving' the street market – which, after all, is what urban renewal is meant to achieve – and how that will affect the making of place and atmospheres.

Ten Katemarkt and De Hallen, side by side, create what Richard Florida defines as authentic urban place:

> Authenticity comes from several aspects of a community – historic buildings, established neighborhoods, a unique music scene, or specific cultural attributes. It especially comes from the mix – from urban grit alongside renovated buildings, from the commingling of young and old, long-time neighborhood characters and yuppies, fashion models and 'bag ladies'.
>
> (Florida, 2002, pp. 294–295)

The risk in this mix is that the balance, unless finely calibrated, will tip in the favour of gentrification, as has happened in London's heritage-listed Brixton Market. What is remarkable about Ten Katemarkt, then, is its indifference, so far, to the global trend of gentrifying markets that, as in the case of Broadway Market, has been cited as crucial to the continued existence of markets. According to Freek Janssens, Amsterdam's neighbourhood markets are being situated by the city's council within that narrative of decline and revival that Nick Dines (2009) detected in his research in London. This narrative is a rationale for the ' "easiest option," which seems to be to privatize the market (a euphemism for getting rid of them)' (Janssens, 2014, p. 99). At the same time, there are a number of remarkable markets that confirm the 'revival' narrative. The Noordermarkt in the Jordaan, a Saturday market, has been in existence since the early seventeenth century. In line with the place-images of twenty-first-century Amsterdam, Noordermarkt has morphed into a farmers' market, which dabbles in bric-à-brac and handmade objects, and is promoted on international lifestyle, travel and food blogs (Khoo, 2012). Another monthly market, the Museum Market, has prime position at tourist ground zero on Museumplein between the Rijksmuseum, Van Gogh Museum, and the much-photographed 'Iamsterdam' sign. Instead of mass-market tourist souvenirs based on Dutch tropes such as clogs, tulips and windmills, the stalls at Museum Market interpret the Amsterdam souvenir through art and craft. Food trucks and buskers round out the 'creative vibe'.

Noordermarkt and Museum Market are indicative of how the Creative City paradigm is being deployed to produce place-images of a globalized Amsterdam (Peck, 2012; Lindner and Meissner, 2015). Creative City discourse incorporates making that is indexical to creative work – a pretty broad church these days (Fuller et al., 2013) – within a managerial framework that is tied to urban development. In doing so, the bureaucratic gatekeepers of

Figure 4.4   Museum Market, Amsterdam

the Creative City define what kind of making is 'creative' and, by extension, who is 'creative' in the urban economy, and anything outside its stretch and influence, including a city's unremarkable places, is considered moribund, unworthy of attention, and ultimately not of cultural value. Jamie Peck points out (2012) that Amsterdam didn't start being 'creative' when it officially adopted Richard Florida's (2002) Creative City prescription; Florida merely identified an umbrella term and apparatus for commodifying creativity. Peck says 'the banal nature of urban creativity strategies in practice is drowned out by the hyperbolic and overstated character of Florida's sales pitch, in which the arrival of the Creative Age takes the form of an unstoppable social revolution' (2005, p. 741). Cities that adopt the Creative City paradigm use the creative industries as a vehicle to draw investment, consumers and workers to an area, but the objectives and outcomes in terms of the impact on urban development are not different from any other form of neo-liberalization of the urban economy.

Creative City-aligned placemaking has the perverse effect of creating less remarkable places that are also less appealing to the so-called creative class on whom they depend as producers and consumers. As Chang and Teo have noted, it is the 'vernacular [that] appeals to cultural tourists and members of the creative class who are attracted to places for their authentic flavour and ambience' (2009, p. 344). I had already come across some of the vendors at

Museum Market the day before at Noordermarkt, and except for the preponderance of images of Amsterdam's canal houses, neither of these markets had a distinctive quality, or atmosphere, that separated them as places from numerous other arts and crafts markets in other globalized cities.

Markets and cultural heritage in Hong Kong

One way of preserving remarkably unremarkable places like Ten Katemarkt is to assign them the category of cultural heritage. Zukin says that acknowledging a city's 'vernacular spaces' as cultural heritage is also a way of acknowledging that 'normal processes of change in all modern societies endanger a community's ritual practices and skills as well as built cultural forms' (2012, p. 281). This recognition enables the decoupling of the social and cultural value of markets from the other types of value they contribute to the urban economy.

In Hong Kong, and in Kowloon and the New Territories especially, local markets are a visible, unavoidable component of the urban fabric. Maurizio Marinelli writes that '[s]treet markets and hawking are an organically constitutive part of Hong Kong's history, culture, and socio-economic development. Since the inception of the colony… street markets have always played an integral role in shaping the landscape for population growth and urban development' (2015, p. 44). They epitomize what I think of as the inside-out quality of place in Hong Kong, where it is as if the small, expensive interior spaces of the built environment cannot possibly contain the volume of making happening within them, and these spaces turn themselves inside out onto the street and public places. Markets are Hong Kong's cluttered outdoor rooms, and making in them produces place that is everyday, public and collective.

Despite their social and material utility, the future of Hong Kong's markets is far from certain in a spatial economy where the scarcity of space encourages an unquenchable demand for it. In Hong Kong, as in many other globalized cities, the exchange value of the space of a market trumps the use value of the place of a market. In the Peel Street and Graham Street market area in April 2015, vendors huddle in front of hoardings covered in images of giant fruit and vegetables. The street scene in one of the last wet markets on Hong Kong Island is as busy as when I visited four years earlier, although the built environment has been thinned out by the removal of stalls and the demolition of shops and residential buildings. The hoardings, in spite of their colourful images, cast an ominous pall. They have been erected by Hong Kong's Urban Renewal Authority (URA) around a multi-site high-rise development incorporating office, hotel and residential space. The URA's plan, which deploys the bland language of regeneration, claims to be retaining the 'vibrancy of the neighbourhood as well as preserving the identity of the place' and producing 'synergy with the current street market' (Peel Street/

Figure 4.5   Inside out

Graham Street Redevelopment Scheme, n.d.), yet contains no future plan for
the wet market in its present form. There is a mention of retail floor space
within the development for selling fresh food, but this is not explicitly identified as space for market stakeholders, who would presumably face rent rises
and increased prices if the market was moved inside. Any direct communication that the vendors have received from the URA about their future has
been as vague as the URA's language on their website (Blake, 2014). While
some of the buildings in the market have been quarantined from demolition

through heritage listing, any businesses who do hang on will undoubtedly be affected by the inevitable changes in demographic and spatial use of the area once the development is completed, and by the fact that they are now located in a heritage area, which puts limitations on everyday use.

A heritage designation can constitute an argument for the value of tangible and intangible culture, and ensure continuity, but the recognition has its own complications. The United Nations Centre for Human Settlements has highlighted the stultifying cultural effect heritage designations can have

Figure 4.6   The last wet market on Hong Kong Island

when they are used to cordon off conservation areas where '[a]uthenticity is paid for, encapsulated, mummified, located and displayed to attract tourists rather than to shelter continuities of tradition or the lives of its historic creators' (quoted in Knox, 2005, p. 4). The dismantling of the Peel Street/Graham Street market is the loss of intangible cultural heritage, which demonstrates how protecting tangible cultural heritage is no guarantee that what animates it, and presumably contributes to its cultural value, will also be protected. Indeed, conservation of the built environment as cultural heritage is often to the detriment of intangible cultural heritage because it leads to gentrification and displacement (Zukin, 2012). Categorizing markets as intangible cultural heritage takes into account the important social role they fill, yet can serve urban development agendas because, unlike tangible heritage in the form of a building, the social can be said to be conserved even if it has been relocated and perhaps rendered meaningless. Anchoring intangible cultural heritage to tangible cultural heritage might be a way of preserving place-based market culture in Hong Kong like the Peel and Graham Street market. However, even twinning the designations is unlikely to have much effect in an urban landscape where cultural heritage, tangible or intangible, is of secondary or no concern in urban development objectives (Barber, 2014).

Down the hill from Peel and Graham Streets is Central Market, a rare example of tangible cultural heritage from Hong Kong's market culture. It is an aesthetically distinctive building built in 1939 in the streamline *moderne* transitional style between art deco and modernism. It is a remarkable space, not only for its singular architectural value, but also as a vestigial space in Hong Kong's built environment. Interestingly, its status as tangible market culture was instrumental in the decline of intangible market culture. As a covered market, it was part of the Hong Kong government's drive to 'turn street markets into indoor public markets…, in the name of progress and modernity, and, of course, for the sake of public health. The outcome has been the development of "modern", more "civilized" and "hygienic" urban spaces, with the collateral damage of the annihilation of "living heritage"' (Marinelli, 2015).

As a residual public space on prime real estate, Central Market is under threat. Up until its closure in 2003, the market underwent an orchestrated decline. Even though it was an apparatus in the removal of hawkers from the street, Central Market was still dependent on itinerant sellers as customers, many of whom bought their produce and wares at the market. Trade was affected by the marginalization of street selling through successive government clampdowns (Marinelli, 2015), and in the 1990s the top level was reconfigured into a shopping arcade and walkway between the Central Mid-Levels Escalator and the Hang Seng Building. Today, a few stalwart kiosks remain in the arcade among the shuttered shopfronts and 'Relocated' signs. Mostly it functions as a very busy thoroughfare between office buildings.

Figure 4.7   Central Market, Hong Kong

A sign at the entrance of the building installed by the URA announces the intentions for the site, which is to construct 'an additional leisure cum entertaining place that can rarely be found in this busy district. It will become a new hang-out for white-collar workers in the daytime, and locals as well as tourists in the evenings and on weekends'. There is no mention made of a market. Inside, there is a 'community consultation' section with diagrams and maps and invitations to the public to share their ideas about regeneration of the building. Like many urban renewal projects in transition, Central Market has been given a Creative City, pop-up makeover. A parklet called 'Central Oasis' has been installed inside, and the URA and Public Art Hong Kong have invited comic artists to cover the exterior with their doodles. The brief for Comics@Central was to 'renew' the exterior with images of 'the hustle and bustle of city life'. In their visual interpretation of this theme, none of the artists referenced the building's history as a market.

Groups campaigning for the building to retain its market heritage have also deployed the language and forms of the Creative City paradigm to argue for Central Market's future viability, significance and relevance as a public place. Citing global examples where markets have been successfully used in place-making and place marketing, Ada Wong (2015), a spokesperson from the resident-activist Central Market Concern Group, proposed an iterative process of conserving market culture that draws on the 'creativity and energy' of La Boqueria in Barcelona and Borough Market in London. Wong suggested that the establishment of a 'refitted market' could 'be co-created and co-shared, to suit 21st century needs', with 'fresh produce stalls and eateries celebrating Hong Kongers' love of food' and 'space for arts and crafts, pop-up lifestyle stalls and co-working clusters, catering to the different needs of 21st century hawkers, entrepreneurs, social enterprises, makers, innovators and creative practitioners' (Wong, 2015).

Wong's research may have led her to another of these vestigial structures, the new creative complex PMQ, which opened in 2015. PMQ, an acronym in the style favoured for urban regeneration projects worldwide, stands for Police Married Quarters and is located on Hollywood Road in the upper section of Central, an expat-favoured and rapidly gentrifying area of Hong Kong. It is a complex of studios, shops, cafés, bars and gardens installed in a nineteenth-century colonial edifice and a handsome 1951 building that used to house married police officers. A brochure available on-site tells us that PMQ is a 'creative industries landmark... thereby creating a platform and instilling power in Hong Kong's creative industries'. A tour group made their way from studio to studio where a lone artist/craftsperson/salesperson sat or rearranged the art/craft/merchandise. Clusters of creative workers sat with their laptops in the café, where a cappuccino costs more than it does in London or Sydney. The complex has that empty, half-finished air that new developments have, an atmosphere bolstered by the deafening sound of pile-drivers from surrounding construction sites. Visitors weren't quite sure what they were meant to be looking at and appeared lost, the phenomenon I had also noticed at De Hallen in Amsterdam. They were placeless. For now, PMQ, an unremarkable use of a remarkable space, is too. Central Market's future, as a market or not, appears equally placeless.

Place and placelessness in Beijing's markets

In *Beijing time*, Michael Dutton visits a rag-pickers' market that is 'off the city map' (2008, p. 142). In contrast, the first three markets on my Beijing itinerary were central, highly visible and well-marked on the tourist map. Each was a different type of market – Donghuamen Night Market sells street food; Xuishui 'Silk Street' Market stocks fabric, clothing and accessories; and Panjiayuan or 'Dirt Market' trades in antiques and curios. None of them could be considered an everyday market in their current form, although according to their recent histories, all originally (re)emerged as an organic part of the post-Reform mercantile landscape. All three markets were presented in promotional material and guidebook entries as remarkable for exhibiting a distinctive Beijing character, yet the goods on sale were not always distinctively local, nor were the marketplaces themselves particularly parochial. They had all been redeveloped in the past decade, clearly with the middle-class tourist and consumer in mind and 'are emblematic of new middle class leisure destinations patronized by urban professionals' (Cartier, 2013, p. 48). Therefore, the conception and presentation of the consumer experience might be considered an articulation of the local in its equivalency with a dominant paradigm of middle-class consumer practices in

China today. The modes of regulation acting upon the spatial and social organization of the markets are also clearly susceptible to local and localized forms of governance and governmentality. Despite movement towards the local, these remarkable markets were ultimately symptomatic of placelessness, 'a process that is, ironically, reinforced as people seek authenticity through professionally designed and commercially constructed spaces and places whose invented traditions, sanitized and simplified symbolism and commercialized heritage all make for convergence rather than spatial identity' (Knox, 2005, p. 4).

The first stop was Donghuamen Night Market in the LED-lit streets of central Dongcheng. From late afternoon, if you turn off Wangfujing Dajie, the pedestrianized shopping avenue whose shops run from Armani to Zara, and head west down Dong'anmen Dajie, you will find a series of permanent stalls of identical size, with standardized signage in Mandarin and English listing set prices for street food from around China. Staff at every stall wear the same red aprons and visors. A booth at the eastern end promises to oversee the health and safety of the market. The man inside is dressed in the militaristic uniform that is *de rigueur* in China for anyone employed to police public conduct and practice. He has his feet up and is reading a newspaper, probably because his domain appears to manage itself. There is very little of the mess and detritus that accumulates on the ground at other food markets. The stalls are immaculately presented, the food neatly displayed. Waste is disposed of discreetly by the vendors, and street sweepers efficiently whisk away any litter that falls to the ground.

Figure 4.8   Donghuamen Street, Beijing

The street sweepers are not the only ones on guard. The market is set up in the fenced-off lane of a busy road. Unlicensed merchants linger on the other side of the fence from the market and undercut the prices charged by the licensed stalls within the demarcated market space. Outside the perimeter, scavengers also eye the diners and wait patiently for them to throw away their rubbish. Avatars of the usually out-of-sight Bajiacun rag-pickers' market reach into the bins and collect recyclable bottles, cans and packaging. They are representatives of Beijing's 20,000 refuse collectors who Dutton describes in *Beijing time*:

> With listless and weathered faces, the Bajiacun ragpickers pedal around town in business suits. With their slightly underweight bodies, their drab and grimy suits, they present a parody of the successful besuited business-man who watch them from their Audis. Indeed, twenty years earlier, the observers could very well have been those observed. From the rear, this traffic looks like a peasant army on bicycles, tricycles, mopeds and on foot.
>
> (2008, p. 148)

On the side of the fence licensed for the formal economy, a man clutches in his fist a bouquet of snakes threaded on skewers. His girlfriend pops some fermented tofu in her mouth with a toothpick. Some of the food on sale, like the tofu, is what you would find elsewhere on the streets of Beijing. Earlier in the day we had seen a woman and her food cart perched on the side of a fre-netic roundabout where she was selling fried buns stuffed with spiced minced meat. At Donghuamen the same buns cost 15 yuan; on the street, a third of the price. A tourist in a fedora hat aims his camera for a food-blogger-style close-up of the chargrilled animal flesh on a stick that he has bought at prices significantly higher than your average street-side stall.

At the end closest to Wangfujing Dajie, from where most of the foot traffic is funnelled, a man asking for spare change sits underneath an illuminated sign. The sign details the market's history, its transformation from a collec-tion of stalls in the 1980s to the tourist destination that it is now:

> In 2000, to carry forward the Chinese culinary culture and enhance the friendly exchanges with foreign countries, the people's government of Dongcheng District rebuilt the night market for dainty snacks with the objective of integrating the traditional delicacies with the modern busi-ness facilities, combining the culinary culture and sightseeing.

I am surprised that Donghuamen has such a short history (as do Panjiayuan and Silk Street), because I had imagined that the night market behind me was the themed version of a traditional marketplace that had occupied the spot for centuries (Miles, 2010). Of course, my assumption makes no sense: the small privately owned businesses that markets support were not possible prior to Reform (Veeck, 2000). I was unable to ascertain whether Donghuamen had been the site for a market prior to 1949, although at nearby Dong An, which has been redeveloped into a festival marketplace and shopping centre, a market first opened in 1902 (Broudehoux, 2004; Cartier, 2013).

The text on the sign emphasizes the market as a space subject to productive government intervention. In the case of Donghuamen it is to ensure food safety and hygiene, and thus to provide a reassuring environment for consumers who might be cautious or anxious about street food, a perhaps misdirected anxiety given recent scandals in China regarding mass-produced food (Veeck et al., 2015). The largest proportion of visitors strolling the market appeared to be middle-class domestic tourists for whom a sanitized, approximated street market experience might offer a more attractive alternative to visiting an actual street market. We saw this phenomenon at work in Qibao too, a canal town on the Metro line just outside Shanghai, where crowds of visitors clung tenaciously to the grid of attractively recreated ancient food stalls that constituted the designated tourist precinct. Step only one street outside the zone and the tourists disappear. In this parallel alleyway, the hole-in-the-wall eateries that sell similar food to local residents for a fraction of the price are deserted.

One is hard-pressed in Beijing to find any unremarkable markets in the central tourist district of the city, where most of the marketplaces have been, to deploy official terminology, 'streamlined' and turned into remarkable places. Regulation is one rationale for the current configuration of the three central markets I visited. The government's narrative on the demolition of the streets of Silk Alley, a textile and clothing market in Chaoyang, and on the subsequent construction of the single edifice 'Silk Street' that replaced it, maintains the necessity of regulating commerce and, in this specific case, the distribution of counterfeit designer goods. Signs, notices on bulletin boards and plaques reiterate this objective, and certainly, knock-offs are not *overtly* exhibited. However, as one walks through the emporium, vendors frequently whisper famous European designer brand names and flash a Louis Vuitton wallet or similar object discreetly folded into a piece of cardboard or paper. On the sole occasion we agree out of curiosity to look, we are led to a locked and alarmed back room where the counterfeit designer brand handbags and accessories are lined up along the wall. Since the redevelopment, the market has continued to be in the news for IP violations, a concession that does not admit the futility of policing piracy, so much as it is designed to illustrate that piracy is being managed by the government. The 2010 arrest of a former manager of the market, which was announced in an English language YouTube video (Tantao News, 2010), focuses on a single perpetrator and is thus a smokescreen obscuring a diffuse, widespread and global informal economy.

The spatial reconfiguration of the market from a series of alleyways to a consolidated six-storey shopping centre is positioned on the website as a continuation of the brand of Silk Street: an evolution that simultaneously incorporates a 'Century-Old shop' and an 'international shopping mall' (Our introduction, n.d.). Even the original Silk Alley was in existence barely 30 years, let alone 100, so this ambivalent appeal to nostalgia has no basis in Silk Street's history, in either iteration, just as the appeal of the street market has been obliterated in all but name in the new edifice, which is more in line with Beijing's place marketing as hub of modernization (Broudehoux, 2007). Silk Street's ambivalence towards the past is clearly an instance of how '[n]ew transnational shopping malls are built on sites of significant historic markets, drawing associated meanings from past places' (Cartier, 2013, p. 52), but it is the modern attributes of convenience, amenity and safety that are underlined as advantages of the new market in promotional material. Of the markets I visited in Beijing, this was the only one that had off-street parking to facilitate the tour groups who are constantly unloaded from buses and channelled through the centre.

The new textile market has diversified and visitors can try Beijing specialities such as Peking Duck, as well as purchase calligraphy, antiques, jewellery, carpets and handicrafts. Silk Street therefore intertwines the globally ubiquitous consumption of brand-labelled goods (regardless of their authenticity) with cultural consumption and cultural tourism. The centre's website also positions the act of bargaining, which is central to most transactions at Silk Street, as a cultural particulate of the city, and as 'one of Beijing's famous scenes' (Our history, n.d.). This is one way to present the ever-shifting and at times uncomfortable power dynamic between well-rehearsed vendors and their unpractised customers, who nonetheless wield leverage in the form of the yuan in their wallets. Effectively, a universal act of capitalist exchange is rearticulated as a local customary act, while also claiming to preserve one of the pleasures of the street market mode of consumption.

Another market on the Beijing tourist trail is Panjiayuan, a flea market that Lonely Planet (McCrohan and Eimer, 2015) touts as one of the capital's top attractions. Panjiayuan initially started out in the early 1980s as an articulation of the ghost market. Dutton explains:

> Ghost markets enabled aristocrats to maintain their social face, while secretly engaging in that most un-Confucian of activities, commerce. Today in China there is no longer any shame attached to commerce, and the ghost market of Panjiayuan no longer masks an aristocratic secret.
>
> (Dutton, 2008, pp. 216–217)

Panjiayuan was a coordinate in an informal economy where the proletariat illicitly sold off the family heirlooms and household treasures before the trade in art and antiquities was legalized in the mid-1990s. As it expanded and gained government recognition and permission, it evolved from an actualised street market located in a Chaoyang *hutong* (laneway) to a site-specific marketplace. The space where the market is housed underwent redevelopment in the mid-2000s in the lead-up to the 2008 Beijing Summer Olympics, and there is none of the haphazard disorganization of accumulated junk that confronts the visitor at a genuine flea market. As with the merchandise, the space is neatly ordered. A hierarchical gradation of traders and commercial spaces begins with a cloth on the ground, and ascends to undercover stalls and, at the top, established shop fronts around the perimeter.

For Dutton, the definitive Panjiayuan narrative is the search for the real, the authentic, the genuine among the endless row of fakes (2008, p. 220). Dutton, borrowing from Walter Benjamin, compares the quest to that of the literal rag-pickers at Bajiacun. Ironically, present-day rag-pickers might be disappointed not at the absence of treasure among the trash, but at the absence of trash among the orderly and meticulous stalls. Nevertheless, Dutton detects at Panjiayuan some of the flea market's promiscuous and indiscriminate meetings of refuse and commodity: 'From traders in antique porcelain to those who trade in the paraphernalia of the Mao years, the significant and the insignificant, the fake and the real, the artistic and the kitsch mix so effectively' (2008, p. 221). The heterogeneity of Panjiayuan's Cultural Revolution memorabilia is where one can rediscover the marketplace's heterotopic rupturing of categorization (Stallybrass and White, 1986) that has so far been elusive in the placelessness of Beijing's markets.

I visited two other markets in Beijing: the city's wholesale fish market and a neighbourhood market in its proximity. Jingshen Seafood Market does not have the same profile as the previous markets because its patrons are mainly local and commercial. It is located in a residential district south of the 3rd Ring Road. Beijing is inland and, unlike coastal cities like Sydney that capitalize upon their proximity to the ocean and therefore a perceived immediacy with the provenance and consumption of seafood, its fish market is not a tourist destination despite seafood being a characteristic feature of China's cuisines. When I asked the staff at the hotel to translate the pinyin address into characters to direct our taxi driver, they told me that they had never heard of it. Information in English language about Jingshen's history is hard to come by, but one piece of information I did glean is that it used to be situated underneath the redeveloped Hong Qiao Pearl Market, a five-storey retail extravaganza opposite the Temple of Heaven in Chongwen. The redevelopment of Hong Qiao was also predicated with the regulatory aim of limiting informal economies and illicit trade.

The infrastructure at Jingshen appears fairly new, if well-worn, and like Panjiayuan, the space within the market is organized with a hierarchy of vendors that goes from those who sell out of the back of vans, to small stalls in the main building and at the top of the heap, larger fitted-out concerns at the edge of the car park. It is well set up for distribution, with a whole section dedicated to packing and shipping. There is not as much care put into presentation at Jingshen as at the branded markets. There is ad hoc invention and bricolage evident everywhere and the resourceful fashioning of tank and filtration systems using materials to hand. Refuse and polystyrene packing piles up in the corners and on the floor. The less polished aspect is due to Jingshen's messy business as a wet market and to the large scale of its operations. (The dried goods section that took up the first floor is much tidier, although under-lit in an attempt to conserve energy; the clerks slumped over their desks or heads thrown back, slack-jawed, are also conserving their energy after the morning rush.) Above all, it is because Jingshen is not instrumental in branding the city.

Down the street from the fish market is the Guancai local market, which carries fruit and veg, meat, dry goods and household wares. Here we are observers of the everyday life of Beijingers in a neighbourhood that is unexceptional in every way – down to the McDonald's being built across the road. If one was looking for a 'real' Beijing market, Jingshen and Guancai might be that, but the very act of outside scrutiny of place causes the everyday to rearrange itself around our presences, however minutely. I doubt Guancai is going to be appropriated by tourists seeking sites/sights where they can be witness to other everydays anytime soon. The aestheticization of the banal that Pierre Bourdieu (1984) observes at work in the construction of middle-class taste succeeds at markets like La Boqueria in Barcelona or Queen Victoria Market in Melbourne. These markets exist before and beyond tourism, yet they simultaneously possess mitigating attributes that attract outside visitors: they are proximate to other centres of tourism; they have architectural merit; they trade in specialist local products. Guancai, on the other hand, is surrounded by dusty, car-choked main roads and empty lots, and sells mass-manufactured pans and plastic shoes.

Figure 4.9 Guancai market, Beijing

There are many other markets like Guancai – in Beijing, in China, in cities throughout the world. In spite of their unremarkable character, their ordinariness, is it possible to regard the market elsewhere with anything other than the tourist gaze as described by John Urry?

> Tourist experiences involve some aspect or element that induces pleasurable experiences which, by comparison with the everyday, are out of the ordinary... There is the seeing of ordinary aspects of social life being undertaken by people in unusual contexts... Visitors have found it particularly interesting to gaze upon the carrying out of domestic tasks... and hence to see that the routines of life are not that unfamiliar... There is the carrying out of familiar tasks or activities within an unusual visual environment... shopping, eating and drinking all have particular significance if they take place against a distinctive visual backcloth. The visual gaze renders extraordinary, activities that otherwise would be mundane and everyday.
>
> (Urry, 2002, pp. 12–13)

Urry's inventory of everyday scenes are ones that are nevertheless designed to be apprehended and consumed within the tourist/tourism complex. What are the consequences for the tourist gaze and for the object of that gaze when one is in a place that is positioned within no other context but the unremarkable? Perhaps this breakdown in ontological demarcation between tourism and the everyday is the reappearance of the liminality that Agnew (1986) identifies as a defining feature of the marketplace, which we had to look hard to find in the placelessness of Beijing's tourist markets but were still able to glimpse in the city's remarkably unremarkable everyday markets.

5 Dis/placing markets through urban renewal

Sydney

The previous chapter explored the challenges facing everyday, unremarkable markets as places in globalized cities. In this chapter, I want to investigate how urban renewal poses particular challenges to markets and place. While urban renewal leads to the loss of place as markets are closed down, relocated, redeveloped or repurposed, other types of markets are being deployed as key apparatus in support of placemaking for urban renewal projects. In a journal issue dedicated to investigating the use of markets as an urban development strategy, Janssens and Sezer write:

> [L]ate capitalist society has paradoxically snatched the romantic image of the marketplace as a tool for urban branding and placemaking. The revitalization that is being pursued in these practices is commonly linked to exclusive private housing and retail projects. In other words, the romantic image of the marketplace serves in many cases as a catalyst for gentrifying neighbourhoods.
>
> (Janssens and Sezer, 2013, p. 169)

Gonzalez and Waley, in their study of a public market in Leeds, also noted 'an emerging interest from the state at different levels in markets as a new regenerated commercial space that brings together various policy areas: urban renaissance, healthy living, community cohesion, urban sustainability, re-localization of the economy and tourism' (2013, p. 965).

In interrogating the connections between markets, place and urban renewal, I want to propose that this relationship is nearly always predicated upon displacement. To explore this proposition I am using the case study of Carriageworks Farmers Market, a market that has been mobilized as an overtly directed vehicle for urban renewal in the inner-city Sydney suburb of Redfern. The displacements associated with Carriageworks Farmers Market are framed by ongoing difficult and 'incommensurable' relationships with place and displacement for Aboriginal and Torres Strait Islander peoples and non-Aboriginal people in Australia (Moreton-Robinson, 2003). Indeed, Christopher Mayes (2014) had submitted that Carriageworks Farmers Market is a continuation of *terra nullius*, which positions displacements accompanying urban renewal in Redfern as historically and spatially contiguous to the dispossession of Aboriginal and Torres Strait Islander peoples in the colonial expropriation of land.

Markets and urban renewal

Urban renewal is a strategic and centralized planning instrument adopted by the public and/or private sector to facilitate an influx of capital into renewed urban centres. Urban renewal has a longer history than gentrification, is usually conducted at a different scale, and involves different types of actors and different relationships to place. Gentrification is 'the transition of inner city neighbourhoods from a status of relative poverty and limited property investment to a state of commodification and reinvestment' (Ley, 2003, p. 2527) through a seemingly organic pattern dictated by individual actors and agents. As the term implies, gentrification is connected to class through its transformation of the social make-up of an area (Glass, 1964), and also in that it is an articulation of taste and cultural capital (Ley, 2003), and ultimately, distinction (Bourdieu, 1984). However, as housing and property in inner cities around the globe have been priced out of the range of all but the wealthiest citizens due to gentrification, that taste-based model has become increasingly obsolete (Tonkiss, 2005). Urban renewal and gentrification overlap with the shared goal of urban development that maximizes and recuperates flows of capital. One can be the foundation for the other. The urban renewal led redevelopment of the Mercat Santa Caterina in Barcelona was conceived as a 'micro-intervention' in a working-class area (Arbaci and Tapada-Berteli, 2012), but the remodelling of an everyday market hall into an architect-designed global food and design destination provided the impetus for gentrification of surrounding streets. At Spitalfields Market in London it was an inverse process. As I outlined in Chapter 2, the gentrification of Spitalfield's housing stock by incoming middle-class residents was incompatible with the operations of the wholesale produce market, and once the market was relocated, its site was tagged for urban renewal (Jacobs, 1996).

There are four main ways in which urban renewal impacts on markets. The first is through the regeneration of an existing market. Improvements to the amenity of the market are designed to attract and service the needs of new consumers and entrepreneurs. The outcome of this is usually that everyday, traditional or general markets are reconfigured as specialist markets. Examples of this that I have already talked about include Mercat Santa Caterina and, in Chapter 2, Broadway Market in London. There is a metonymical logic at work here: the regeneration of the market is the regeneration of the area. The second way is to establish a brand new market as a gentrifying or placemaking device in support of an urban renewal project. Carriageworks Farmers Market, the case study in this chapter, is an example of this. The third way is to repurpose residual infrastructure from markets that have been relocated or closed down because they have 'outgrown' their site, their trade has become obsolete or unsuited to a transforming area, or the local catchment community has undergone demographic change. Often this repurposing retains inter-textual references to the market. For instance, the entrance to the University of Technology Sydney library, which is housed in a former market hall, is flanked by blown-up archival images of the market it replaced. The fourth is the demolition of a market altogether, replacing it with something different, as happened in Rio de Janeiro in the 1960s when its grand Municipal Market was torn down to make way for an elevated roadway.

Recent models of urban renewal in globalized cities have eschewed the latter model of top-down, large-scale destruction of extant infrastructure, in favour of capitalizing on its heritage and aesthetic value, much in the way that gentrification has in the past. As Jane M. Jacobs has observed:

> Contemporary urban transformation is far more likely to engage consciously with the local character of an area than rapaciously obliterate it. This is perhaps most clearly seen in the varying ways in which heritage is mobilized as part of the legitimating framework of urban transformation.
>
> (Jacobs, 1996, p. 72)

Place that is encapsulated by the convenient catch-all of 'heritage' has a neatness to it because it can be relegated to the past, and can therefore be reduced to the symbolic or representational without the messiness of the material present. Place is still alluded to for cultural authority or authenticity through the tokenistic metonym: public art depicting portraits and scenes of previous communities; blue plaques; a carefully curated piece of renovated industrial detritus. Urban renewal (more so than gentrification) frequently requires that pre-existing manifestations of place that do not substantiate the place-image (Shields, 1991) of the transformed urban landscape are, at the least, managed and, at worst, marginalized and even eradicated. The devaluing of place functions as the justification for the displacement of existing communities (Holgersson, 2014). As globalization and urban transformation undermine or remove place, placemaking seeks to re-instate it. Sharon Zukin notes that the appeal to heritage constitutes a form of placemaking where place has most likely been scrubbed away: 'At best, when market forces destroy and re-create an existing landscape, its artefacts... are stored, restored, and even relocated to create an "authentic" sense of place' (1991, p. 20).

The same consideration for 'preservation' and 'conservation' is rarely shown to the existing communities who are being displaced. Nick Dines (2009) witnessed a displacement scenario at work in the proposed redevelopment of Queens Market in East London. Even though those who used the market viewed it as a lively and welcoming place, the lexicon of urban blight was deployed by local council and prospective developers to demonstrate a need to privatize the space. The council's plan side-lined a shopping precinct and meeting place that was well-utilized by the community in favour of a supermarket and residential development. As in Broadway Market, diverse groups who used Queens Market were drawn together to save it from a scheme that eschewed improvements to existing market infrastructure, and instead replaced it with something designed for imagined future residents. This, Dines argues, is not an uncommon situation in the UK.

> During the last two decades there has been a narrative about the decline of markets. Traditional markets have found themselves closed down, under threat or relocated outside urban centres, largely as a result of growing competition from superstores and out-of-town shopping malls and a lack of investment from local authorities which have redirected finances towards higher priorities such as housing and education (Watson and Studdert, 2006). However, the idea of 'decline' is at the same time problematic. Conceived simply as a drop in customer footfall, it over-looks the ongoing, less tangible social role of markets.
>
> (Dines, 2009, p. 257)

These narratives of decline are strategic because, as Neil Smith (1996) points out, labelling urban areas as degraded or devitalized provides the rationale for intervention in the form of the urban renewal project.

A narrative of decline also justifies the perception of urban space being 'up for grabs'. The vacant buildings, workshops and factories left behind in the post-industrial landscapes of twentieth-century inner cities enhanced the notion that this was legitimate terrain for gentrification and urban renewal. It confirms the self-image that gentrifiers have of being 'pioneers' entering unsettled areas (Metro US, 2015; Tonkiss, 2005). Rebecca Solnit writing about urban renewal in San Francisco talks about the city being 'delivered vacant' to developers (Solnit and Schwartzenberg, 2002). By this, Solnit means that poor, minority and long-term residents were moved along and out by rising rents to smooth the way for urban transformation and to obscure the accompanying displacements. Mike Lydon and Anthony Garcia are principals of the Street Plans Collaborative, a consultancy at the centre of the global placemaking industry whose motto is 'Better Streets, Better Places'. Lydon and Garcia understand their task to be the activation of 'vacant lots, empty storefronts, overly wide streets, highway underpasses, surface parking lots, and other underused public spaces' (2015, p. 6). They read these spaces in the city as a vacant terrain waiting to be turned into place by placemaking processes. This, they call tactical urbanism and they claim that '[o]pportunities to apply Tactical Urbanism are everywhere – from a blank wall, to an overly wide street, to an underused parking lot or vacant property' (2015, p. 87).

Sydney's contemporary markets

I have lived in Sydney for most of my life, a city whose entire urban devel-opment has been based on the fiction that the land on which it was built was vacant. I have early memories of its markets. I remember going to

Paddy's Markets when it was still held in a market hall at Haymarket in the city's centre. Later, when I was in high school, it was second-hand markets with my girlfriends on a Saturday. Either Balmain Market, where the stalls were placed higgledy-piggledy in the uneven grounds of a small church-yard, or the hectic and socially intimidating markets in Paddington, a buzz-ing gentrified suburb that was the epicentre of fashionable 1980s Sydney. I swore off Paddington Markets in the early 1990s, vowing never to go back because it was so crowded, expensive and pretentious. The retail landscape in Paddington has since undergone a dramatic reversal in fortune, with many of the designer boutiques along its main street closing down after a large Westfield shopping centre was opened in nearby Bondi Junction in 2003. Recently, I returned to Paddington Markets for the first time in more than two decades, and the market was a much more pleasant and manage-able affair in terms of size, atmosphere and crowds. In the 1990s, Glebe, Rozelle and Bondi Beach became my favourite markets, not only for buy-ing, but for making a bit of extra money from selling my pre-loved clothes and bric-à-brac.

I am still a regular market-goer. A Saturday farmers' market started up the road from us earlier this year. We walk there to do our weekly fruit and veg shop because it is more convenient than driving to the shopping centre and the produce is fresher and no more expensive. On Sundays in the suburb next to ours is Kingsford Rotary Markets, a true flea market where sellers throw junk on the asphalt for buyers to sift through. The prices are low and the crowd is mixed. There is no one buying or selling expensive mid-century lamps or handmade tote bags. Kingsford Rotary Markets are organized by a local charity organization and operate out of a council car park, the kind of under-capitalized public space that is endangered in the globalized city. Recently, the local council has proposed building a multi-storey car park on the site, which will undoubtedly impact on the market's operations (Randwick City Council, 2014).

Sydneysiders like markets too, if the new markets popping up each month are an indication. A survey taken in August 2015 of one website dedicated to market listings (localmarketguide.com.au) revealed 65 weekly and monthly markets in Sydney. It is difficult to obtain reliable numbers on Sydney's markets because there are few centrally collected statistics and figures, or associations representing market operators and vendors. The Australian Farmers' Markets Association is a small body, but there are no industry-wide national organizations such as in the UK, where the aligned National Market Traders Federation and National Association of British Market Authorities represent market workers and managers respectively. There are also few secondary sources on the spatial, social and cultural role of markets in Sydney's spatial and social development. The secondary sources that do exist tend to be tourist guides or write-ups in lifestyle media. Michael Christie's (1988) history of Sydney's central produce market is a rare example.

The absence of sector-wide representation is surely a symptom of Sydney's atomized market culture, as is the absence of data. The dearth of histories or cultural commentary on markets, or on their role in Sydney's consumer culture (Crawford et al., 2010; Kingston, 1994), is further confirmation of the marginal position to which they have been relegated in the retail landscape of Sydney. The city's markets rarely have dedicated infrastructure and, as in the case of Kingsford Rotary Markets, are frequently run by community groups and charities as small-scale, temporary affairs out of multi-purpose sites such as schools, churches or car parks. The operation of markets in Sydney is regulated according to the prioritized demands of other stakeholders in Sydney's urban landscape. Markets are often conceived, managed and promoted as a supplementary option, as opposed to an alternative or parallel mode of shopping and consumption. One example is Bondi Junction Village Markets, a small collection of stalls in a transport, retail and services hub near Sydney's iconic Bondi Beach. The shopping precinct is dominated by the aforementioned Westfield shopping centre that decimated Paddington's main commercial street. According to its own figures, Westfield Bondi Junction has 508 shops, 127,736m^2 of gross, lettable space, more than 21 million visitors a year and over AUD$1 billion in annual retail sales (Bondi Junction: key stats and figures, n.d.).

In its marketing, the Bondi Junction Village Markets highlights its proximity to a transport hub and its utility to commuters and tourists.

> Bondi Junction Markets are nestled under cover at the western end of the Oxford Street Mall just a hop, skip and jump from the Bondi Junction train and bus interchange. The Bondi Junction Village Markets offer commuters, workers, local residents and tourists a bevy of fresh produce, gourmet goodies and artisan crafts each Wednesday, Thursday & Friday.
> (Bondi Junction Village Markets, 2015)

The market is not conceived as infrastructure supporting the everyday needs of a regular, emplaced community. Rather, it is a transient experience for visitors. Like the majority of the 65 markets listed on localmarketguide.com, Bondi Junction Village Markets is a specialist market selling an assigned range of products. Its selection of stalls is carefully curated to meet expectations about the spectrum of goods one might find at a market: organic vegetables, artisanal bread, hippy clothes, imported jewellery and hipster-style craft. Instead of offering everyday shoppers a serious, affordable alternative to the shopping centres with which it is surrounded, it is a placemaking device that deploys consumption as a way of encouraging people to visit and linger in a public space that is usually used solely as a thoroughfare. There is nothing very 'villagey' about the built or social environment in which the market is situated, or the atmosphere it creates.

A short history of markets in Sydney

The emergence of specialist markets in Sydney is partially the result of the centrally concentrated market system that evolved out of the city's colonial settlement. In the absence of a diffusely distributed network of municipal markets, a consumer ecology has emerged in Sydney where privately operated, niche markets that are spaces for consumption (Miles, 2010) have flourished. Markets as they are understood in this book did not exist in Australia prior to colonization. Aboriginal nations exchanged goods, material culture and knowledge (Gammage, 2011), and this trade was conducted along geographical routes that were later appropriated by Europeans in the economic colonization of Australia (Kerwin, 2010). Trade with Aboriginal nations, or even amongst themselves, was not an initial priority for the British when they established the penal colony of Sydney in 1788. Sydney was founded as a militarized prison camp where power was concentrated in the army, and the distribution of goods was a mode of social control (Kingston, 1994). The institutions that developed in the first years of the colony had a shared objective of centralized power, and were not conducive to the establishment of markets or even a market economy (Christie, 1988).

However, the monopoly of government farms and stores was soon devolved to smaller landholders (usually ex-convicts) in order to build food security. In 1792, the distribution and exchange of goods was liberalized with an official decree that farmers were 'at liberty to dispose of such live stock [*sic*], corn, grain, or vegetables, which they might raise, as they found convenient to themselves', and any surplus 'would be purchased by the commissary on the public account at a fair market price' (in Christie, 1988, p. 24). This order, and the arrival of the first international merchants in the same year, resulted in the emergence of the colony's first informal markets at wharves in Sydney and in the upriver agricultural outpost of Parramatta. Regulation of markets in Sydney was formalized in 1806 when the governor of the colony issued a general order in response to black-market profiteering after a recent flood:

> it is ordered that in future no Purchase shall be made until every little thing landed at the Place now appointed; and that the Market shall not be considered to have opened until Seven o'clock in the Morning. That said Market Place shall extend from the End Paling of Daniel McKay's Garden, in the middle of High Street, towards the Parade.
>
> (in Christie, 1988, p. 36)

The relocation of the market moved it closer to government buildings at the centre of the town, as was common in European cities (Calabi, 2004; Slater, 2014). With regulation came restrictions upon trade practices, and rulings that outlawed the practices of forestalling (trade along supply routes) and resale for a profit within four miles of the market. Social behaviour was also policed through ordinances that, for example, limited access to market facilities outside permitted times, determined the proper and improper use of market spaces, and even forbade the circulation of rumours (Christie, 1988, pp. 43–44). This last order is noteworthy as an attempt to stamp out what anthropologist Clifford Geertz has advanced as a fundamental component of a market. Geertz famously wrote that,

> in the bazaar information is poor, scarce, maldistributed, inefficiently communicated and intensely valued… The level of ignorance about everything from product quality and going prices to market possibilities and production costs is very high, and much of the way in which the bazaar functions can be interpreted as an attempt to reduce such ignorance for someone, increase it for someone, or defend someone against it.
>
> (Geertz, 1978, p. 29)

Despite the strict limitations placed upon the sale and distribution of produce and on conduct, by the early nineteenth century the marketplace in Sydney was, according to Christie, 'both the economic and the social hub of the colony' (1988, p. 40), and the market had developed a culture and a sociality that emulated markets in Europe. Contrary to the marginal role to which markets have been consigned in contemporary Sydney's retail and consumer landscape, Christie's history draws upon large amounts of archival documents and images that record how Sydney's material infrastructure was being shaped and built in accordance with the demands of its centralized market operations. A second wharf, Market Wharf, was constructed in 1811 to handle the growing amount of produce coming in from Parramatta (Christie, 1988, p. 45), and a road (which is still named Market Street today) was laid directly from the wharf to George Street, where the market had moved in 1810.

The new George Street location was at that time on the outskirts of town, which had the dual benefits of expediting distribution of goods from further afield, and of reducing disruption to non-market business in the town (Christie, 1988, p. 44). This move was merely the first time that the market would be subject to redevelopment and urban renewal in response to the flows of urbanization. An elegant new structure was completed at the George Street site in 1834, and less urbane business that was 'unsuited' to the new buildings, such as the hay and corn exchange and the second-hand Saturday market known as Paddy's Market, moved south of the city to what is still known today as Haymarket (Christie, 1988). They were joined by sellers who could not afford the increased rents at the redeveloped George Street. This movement instigated urban renewal at Haymarket too, as the yards and abattoirs of its livestock market were replaced by a growers' market, and a new complex, the Belmore buildings, was opened by the city's council in 1869 to accommodate its overflows.

By the end of the nineteenth century, Sydney's two market precincts were compromising the amenity of an expanding civic centre. The volume and the type of traffic produced by the George Street market were not suited to the city's day-to-day business. In 1898, the building was replaced by the Queen Victoria Market, a technologically advanced shopping arcade that was a market in name only, and the 'real markets… moved south [to Haymarket] and would remain there for half a century until the same factors that forced their relocation in 1909 would once more apply and the markets would move again' (Christie, 1988, p. 89). In 1975, the wholesale market was relocated 15km west of Haymarket to Flemington on the rationale that the dirt, noise and congestion that it created were incompatible with current uses of the area, a rehearsal of complaints about the markets that had occurred since before their existence was formalized in 1806. The subsequent sale of the Haymarket site in the late 1980s for redevelopment as retail, residential, leisure and educational space was also determined by the increased value of the centrally located land, and in this sense, is part of a global phenomenon relocating markets to remote urban locations so as to capitalize on their valuable central sites (Duru, 2014).

The flows to and from the central Sydney market shaped the city in other ways. In a racist local labour market that privileged white workers, the market trade provided employment for large numbers of Chinese migrants arriving in the mid-nineteenth century. Businesses run by Chinese merchants, and servicing the requirements of the Chinese market gardeners who 'routinely spent some nights living at their gardens and the rest in the city in lodgings near the markets' coalesced in nearby streets (Fitzgerald, 1997, p. 88), forming Sydney's Chinatown. Sydney's local government system is also inherited from its central markets. In 1839, responsibility for Sydney's market was divested from the governor by virtue of the Market Commission Act. The resulting market commissions were the forerunners of Sydney's current municipal government system. When the City of Sydney was incorporated in 1842, writes Christie, its

> elections were modeled on the market commission… [C]ontrol of the markets was transferred directly… to the new councillors of the Corporation of Sydney, who immediately set about obtaining the valuable land on which the markets were sited. The connection between Sydney Council and the markets was close from the outset. The meeting to elect the first six aldermen took place in the market buildings and one of their earliest actions was to appoint a market committee to draw up a set of by-laws for the running of the markets.
>
> (Christie, 1988, p. 56)

More than two centuries of state control of Sydney's central produce markets came to end in 1997 when they were privatized. Histories of markets emphasize the myriad and arcane overlapping legal and bureaucratic structures that brought the simultaneous and/or contiguous development of markets and cities under centralized state control (Braudel, 1982; Casson and Lee, 2011; Tangires, 2008), yet the particular exigencies of supply and demand in a remote colony ensured a level of centralization in Sydney that is striking. The historical, spatial and cultural consequence of Sydney's intensely centralized market network was that market infrastructure beyond the central wholesale markets was not developed. There was no state policy of building or supporting satellite local markets. As Sydney sprawled, it was grocery stores, supermarkets, chain stores, shopping centres and now specialist markets that met social and consumer demand (Kingston, 1994; Bailey, 2010). In 1971, the founder of Westfield, Frank Lowy, equated the modern mall with markets not merely in terms of its function but in terms of the phenomenology of place. He said that 'Shopping centres are essentially market places', and that 'The shopping centre makes it possible for all who enter its exciting atmosphere to participate… to share an experience… through the visual, aural and touch senses' (2006, p. 281). For Lowy, the shopping centre was not only the technological and teleological evolution of the marketplace, it *was* the modern marketplace.

Placing Redfern

Carriageworks Farmers Market is the type of specialist market that has emerged in Sydney in the absence of a system of subsidiary municipal markets. The market is held every Saturday in the Carriageworks creative precinct, which is next to the historically significant inner-city suburb of Redfern. Redfern has multiple, contested histories of place, and has been shaped spatially and socially by local and global flows. It is Gadigal land in the Eora nation, and in the years since European settlement it has hosted, and continues to host, a politically active and visible Aboriginal community; a number of migrant groups; large populations of working-class residents and public housing tenants; artists; a new wave of creative workers; and growing numbers of middle class and aspirational gentrifiers.

Before it became an arts centre, Carriageworks was a workshop in the Eveleigh Railyards. As Kay Anderson explains, the railyards were one of the few places of employment open to the large numbers of Aboriginal people who migrated to Sydney from rural areas after World War II:

> New assimilation policies designed to foster the integration of Aborigines into European society were implemented nationwide, and these, together with rural recession in New South Wales, forced other Aborigines to Sydney during the 1950s. Inner Sydney suburbs within easy reach of Central Railway station became a magnet to Aborigines of diverse communal and country origins who sought cheap housing, access to public transport, and unskilled employment in the Eveleigh Railway Yards and other industrial outlets.
>
> (1993, pp. 318–319)

This led to Redfern becoming home to the largest community of Aboriginal people in one place at any one time. Gary Foley (2012) estimates numbers to be at between 25,000 and 30,000 people at its peak from the 1950s to the 1970s, a concentration that fostered solidarity, activism and political consciousness (Foley, 2001; Morgan, 2012). After Aboriginal and Torres Strait Islander peoples obtained the vote and inclusion in the national census in 1967 through popular referendum, Redfern was the site for other important steps in the burgeoning struggle for self-determination, including the establishment of the Aboriginal Housing Company (AHC) in 1973. The AHC was the first housing collective in Australia. It was established to circumvent the threat of urban renewal in Redfern, which was set to displace Aboriginal people from Redfern once again, as well as to achieve autonomy from a private housing market that exploited and discriminated against Aboriginal tenants. With the support of the federal government, the AHC purchased and managed residential properties in an area of Redfern that became known as The Block (Shaw, 2007).

The histories of the AHC and The Block are only two of many narratives and experiences of place in Redfern that make it politically, socially and historically significant for Aboriginal and Torres Strait Islander peoples. Larissa Behrendt explains that

> in urban areas, there is a newer imprint and history, one that is meaningful and creates a sense of belonging within Aboriginal communities that have formed in urban areas. This is a cultural and political history that is implanted in the area where we now live. I... think of places such as the Redfern Medical Centre where important community meetings have taken place. Or South Sydney Leagues Club, which attracted young Aboriginal men from across the state, including my uncle, to come to the city and play football. I think of Redfern Park where I heard the then prime minister of Australia, Paul Keating, acknowledge that this is an invaded country.
>
> (Behrendt, 2006, p. 7)

Prime Minister Keating's 1992 visit and speech, to which Behrendt refers above, demonstrated that Redfern's significance as place to Aboriginal and Torres Strait Islander peoples has been symbolically, if not materially, recognized by non-Aboriginal Australia.

Displacing Redfern

Fanning out from Sydney's Central Business District, gentrification took in adjacent working-class suburbs, but Redfern, George Morgan writes, 'remains disjointed from the remainder of this region, a maverick, resistant, and scruffy presence... While other comparable inner urban areas of Sydney have been comprehensively gentrified [Redfern] remains transitional' (2012, p. 211). Its proximity to the city thwarted the gentrification pressing on its borders because its deprivation and poverty were all too visible (Shaw, 2007). This is part of the dialectical movement between visibility and invisibility that Behrendt sees in the treatment of Aboriginal people in media and political discourse:

> media attention becomes intense only when there are socioeconomic problems or racial tensions, such as the so-called 'Redfern Riots' [in 2004].

> It is through these images and stories of youths committing violence,
> engaging in criminal activity and antisocial, self-destructive behaviour
> that the Indigenous presence often breaks in to the consciousness of
> Sydney residents. [...] These images also reinforce the impression that no
> cohesive Aboriginal community exists in urban areas, so we once again
> become invisible.
>
> (Behrendt, 2006, pp. 7–8)

Conspicuously surrounded by the increasingly affluent suburbs of inner-city
Sydney, The Block (which had become a metonym in the popular imagination
for the entire suburb of Redfern) was, according to one media report, causing
a 'blockage' in gentrification (ABC, 2004). This perception was disputed by
chief executive of the AHC, Mick Mundine:

> For too long our community has been a no-go, our streets just a fast track
> to the airport rather than somewhere to stop and learn. For too long our
> residents have been stuck in a vicious cycle of drugs, alcohol, violence.
> Mate, I'm sick of The Block being called The Blockage.
>
> (Mundine, 2008)

However, numerous interested bodies, including the AHC itself, appropriated the discursive presentation of social and spatial impediments to urban
development in Redfern as a means to kick-start urban renewal plans.
Housing in The Block was demolished on the basis that it was unfit for
habitation, thereby creating a space in which the AHC could carry out
redevelopment plans of its own.

At the government level, the New South Wales (NSW) state government
set up the Redfern–Waterloo Authority (RWA) in 2004 to take over planning controls in the area from the local council. The RWA's explicitly stated
remit was to '[revitalize] Redfern, Waterloo, Eveleigh and Darlington through
urban renewal, improved human services and job creation' (Redfern–Waterloo
Authority, n.d., p. 1; see also Morgan, 2012), a remarkably anodyne description
for the large-scale changes that its urban renewal projects had planned for the
social and spatial landscape. Redfern has been subject to large-scale top-down
urban renewal before. In 1946–1949, the NSW Housing Commission declared
the area, along with adjoining suburb Waterloo, suitable for slum clearance. 606 terraces and cottages and more than 30 other buildings were cleared
(City of Sydney, 2011a, p. 16). This approach is what Morgan describes as the
urban renewal of 'heavy-handed modernism, with its ambition for erasure of
residual communities, buildings and landscapes' (2012, p. 212). He advances
that in Redfern (as has happened in other globalized cities) this type of urban
renewal has been replaced with 'the creative cities model that promotes local
conservation and a more organic process of transition' (2012, p. 212).

However, the plans for Redfern from the RWA (which was wound up in
2011) and its successor, the Urban Growth Development Corporation, could

be interpreted as substantiating the racist perception played out in the real estate market that the continued presence of an Aboriginal community in Redfern negatively affects property values or, at the very least, slows their growth. In overlooking the fact that Redfern is, and has been, the base for a number of Aboriginal arts companies (Johnson, 2014) and administrating bodies whose cultural capital is equal to that of any gentrifier, discourses of gentrification and urban renewal reduce the social and material complexities of Redfern's Aboriginal history to market indicators. Few are as candid as DeiCorp, a developer who promoted their 'DeiCota' development in Redfern by referencing race and using racist language: 'DeiCota has good rental return and convenient location. The Aboriginals have already moved out, now Redfern is the last virgin suburb close to city, it will have great potential for the capital growth in the near future [*sic*]' (Karvelas and Rushton, 2014).

The resulting public outrage to this line of marketing was predictable, but by focusing attention on a supposedly extraordinary act of racism, the endemic racism of urban renewal as a re-enactment of the legal fiction of *terra nullius* in Redfern is masked. This outrage is a strategy of what Alana Lentin calls neo-liberal 'post-racialism' where 'persistent distancing from racism' places racism 'squarely in the domain of the historical and/or the pathological' (2011, n.p.). DeiCorp quickly removed the offending text from their website, yet the company's relationship to race and urban renewal in Redfern is convoluted in that it has been contracted to build the AHC's urban renewal project for The Block. The first stage of AHC's Pemulwuy Project prioritizes student housing and retail as a means of subsidizing affordable housing for Aboriginal residents, which will be built later based on returns from the commercial segment of the development (AHC, n.d.). The prioritizing of non-Aboriginal housing and retail over housing for Aboriginal residents who were displaced when housing stock was demolished on Eveleigh Street mobilized an ongoing protest by the Redfern Aboriginal Tent Embassy (see Foley et al., 2014, for a history of the Tent Embassy and Aboriginal protest). The protesters continuously occupied the development site from May 2014 until August 2015, when they secured assurances from the AHC and federal politicians that the construction of Aboriginal housing would be adequately funded.

Rebranding Eveleigh

There are historical precedents for a market at Eveleigh Railyards, which, interestingly, were not unconnected to urban renewal. The advent of railway transportation for food supplies coming into Sydney meant that the hay and grain exchanges moved from logistically unsuited Haymarket to the more convenient railyards in the 1860s (Christie, 1988). In 1989, Paddy's Markets were temporarily housed at the railyards while their central Sydney site was being redeveloped. Carriageworks Farmers Market was inaugurated under the name of Eveleigh Farmers' Market in 2009 as a RWA initiative. In addition to sharing its name with the railyards, the market shared its name with

Eveleigh Street, one of the four streets in The Block. The divide between the Eveleigh brand at the market and this other Eveleigh a couple of streets away appeared unbridgeable up until recently. Due to images and stories disseminated by the mainstream media and perpetuated by hegemonic political discourse, The Block became shorthand in the popular imagination for urban blight and social dysfunction (Shaw, 2007, 2013). This image was amplified by saturation coverage of protests in the area in 2004 that erupted in the aftermath of the death of T. J. Hickey, a young Aboriginal man who died while being chased by police. Visitors to Carriageworks Farmers Market who arrive by public transport at Redfern railway station, one of the focal points of the 2004 uprising, need not make the short walk through the neighbourhood at all; on market days, a shuttle bus runs regularly between market and station, and consumers can negotiate the intervening streets in a moving vehicle.

On a Saturday morning, the view from the concrete ramp that leads down to the regenerated Eveleigh train yards is of a marketplace thick with shopping trolleys, prams and pugs on leashes. Celebrity chef Kylie Kwong, whose reputation is built on her insistence on the local and organic, is flat-out making dumplings and savoury Vietnamese pancakes. In what looks like a parody of a French street market, shoppers in striped tops fill Provençal-style straw baskets with expensive fruit and veg. At street level, members of Socialist Alliance hand out leaflets urging consumers to 'Take the power back!' and an animal welfare organization seeks homes for rescued cats as if offering absolution for the conspicuous consumption down below.

Markets, like cities, are typically messy – materially, socially, spatially. However, in the creative precinct of Carriageworks, the market is almost ironic in its neatness. The repurposed shed that houses the market is post-industrial chic, and it and the market have been used as the backdrop for television advertisements and cooking shows. A cross-promotional video embedded in the market's website features the host from *River Cottage Australia*, the Australian version of the popular British farm-to-table cooking and lifestyle television program (SubaruAustralia, 2014). This confluence of marketing for the show, the farmers' market and a Japanese car maker is an intriguing performance of gender, nationalism, food and the politics of the local. It reminds us that farmers' markets as a node of ethical consumption are about individual technologies of the self, articulated through cultural capital and the communication of taste, as much as any political decision to withdraw from mainstream consumption.

Situating a farmers' market at Eveleigh transmitted messages about the RWA's aspirations for the type of residents, businesses and visitors it wanted to attract to the area. The narrative of place created by the market's rebranding of Eveleigh had the effect of effacing Redfern's and the Eveleigh Railyards' Aboriginal histories. The effacement has a certain irony given that the market trade, along with the railyards, was one of the few sources of employment open to Aboriginal workers in Sydney in the nineteenth and twentieth centuries (Christie, 1988). The market's placemaking strategy reached its apotheosis

Figure 5.1   Carriageworks Farmers' Market, Sydney

when its name was changed in August 2015 from Eveleigh Farmers' Market to Carriageworks Farmers Market, a moniker which escapes the Eveleigh association altogether.

Creative Redfern

As one place-image of Redfern supersedes another, Morgan (2012) observes that expressions of Aboriginal identity are welcomed in urban renewal discourse, but largely those that communicate alliance with the creative economy. An occasional recognition of the symbolic and lived importance of Redfern to the contemporary Aboriginal urban experience is incorporated through metonyms of Aboriginal culture that are complementary to the new place-image of Eveleigh; an Aboriginal arts festival or an informative plaque detailing previous spatial practices. Everyday iterations of Aboriginal identity in Redfern are still overlooked and prevented in decisions about urban planning and design (Behrendt, 2006).

Up until 2014, a monthly 'Artisans' Market' was also held at Carriageworks, which further mobilized the creative industries in the rebranding of the site. In the figure of the artisan, a nostalgic history of working-class, industrial labour merged with a romanticized narrative about the unalienated post-industrial labour of the craftsperson. The recent renaming of the farmers' market as Carriageworks Farmers Market continues this alignment with Creative City discourse (in addition to acknowledging Carriageworks' responsibility for the market, which they took over in 2013). In August 2015, a 'creative director' for the market was appointed who made the following statement:

> Carriageworks is the ultimate space in Sydney... My goal is to put the Carriageworks Farmers Market (one of Sydney's absolute treasures) on the world stage, promoting local produce and creating an internationally acclaimed food and arts precinct in Sydney's Redfern... We're aspiring

Figure 5.2   The Block, Redfern, Sydney

to grow the Carriageworks Farmers Markets to the same international acclaim as London's Borough Markets.

The vision is to propel the existing, already excellent, markets at Carriageworks upwards and outwards to be Australia's best – and on equal standing to the Carriageworks acclaimed Artistic Program.

(McEnearny, n.d.)

The latest rebranding, the appointment of a creative director, and references to Borough Market fully situate Carriageworks Farmers Market within a globalized Creative City discourse (Mould, 2015). As the place of Carriageworks becomes more visible, Redfern becomes less visible, and in this sense the place that is being made in the market is metonymical to place as it is being transformed in Redfern. Increasingly, the right to the city in Redfern means conforming to rhetorics and practices of creativity, entrepreneurship and consumption of the sort that are being produced through Carriageworks Farmers Market. In this formulation, urban renewal conceived through and actualized by the creative economy is dynamic and forward-looking, while 'long-term residents and workers are viewed as residual and culturally inert, tied to the declining occupations and ways of life of the Fordist era. They can only ever form a passive backdrop to the tectonic forces of urban change' (Morgan, 2012, p. 208).

The contemporary conditions of Redfern's spatial economy emerge from a constellation of urban renewal and gentrification processes, contemporary class compositions of urban creative labour, and the competing community interests focused in and on that area. One might suggest that consumers who want to shop locally and ethically should be served by a local neighbourhood market that services their tastes and practices. As Mayes points out, a farmers' market like Carriageworks 'can and does serve to define a community. However, in the process it also excludes those that do not or cannot fit into that imaginary. An implicit exclusion that occurs through community building is not necessarily bad or problematic, but it requires acknowledgment' (2014, p. 280). This is what is problematic about place as is being elaborated at Carriageworks Farmers Market: there is a clear asymmetry between the consumer's investment in the farmers' market as ethical practice, and the evident lack of spatial and social justice. Carriageworks' clientele is one with a high disposable income, judging from the prices being charged for organic, home-grown and handmade produce. It is financially demanding to do one's weekly shop here, and impractical too, given that you would have to travel elsewhere to buy dry goods and other household durables. Nevertheless, proximity to the University of Sydney, one of Australia's prestigious 'sandstone' universities, and the swathes of affluence in previously working-class suburbs such as Newtown and Alexandria ensure a dedicated local crowd and atmosphere. The atmosphere at this market reflects the social and spatial monoculturalism of the gentrifying inner-city streets that surround it.

The visual rhetoric and place-image at Carriageworks Farmers Market is based on the nostalgic parochialism of a village market, yet the emphasis on place through *placing* the provenance of the produce obscures a series of displacements at the site, beginning with the dispossession of Aboriginal and Torres Strait Islander peoples by European settlement. Displacement is inscribed in the very name Eveleigh, which is named after a British colonizer. Writing about Carriageworks Farmers Market, Mayes says 'there needs to be a greater awareness that these spaces are not *terra nullius* and that community development and building cannot be done with complete ignorance or disregard of the communities that existed prior to farmers markets' (2014, p. 282). Behrendt has advanced that the logic of *terra nullius* that was used to justify the violent, racist colonization of Australia continues to apply in Redfern, even as previous displacement (which *terra nullius* denies the possibility of) is used as a means to rationalize current displacements:

> There is also a view that those Aboriginal people who live within a metropolis such as Sydney are displaced, and therefore do not have special ties there. This view can persist even if the Aboriginal families concerned have been living there longer than the observer's family. While it is true that an Aboriginal person's traditional land has fundamental importance, it is also true that post-invasion history and experience have created additional layers of memory and significance that relate to other parts of the country.
>
> (Behrendt, 2006, p. 6)

In late 2014, another market started in the area. The monthly Redfern Night Market at the Redfern Community Centre describes itself as 'a positive space – highlighting Redfern's creative and individual flair, also showcasing local and emerging artists/designers' (Redfern Night Markets, n.d.). It has a focus on craft, street food and music, and classifies itself as social enterprise. The market cultivates a place-image of Redfern as a centre for urban Aboriginal identity filtered through the Creative City. On a chilly Friday night in winter, a reggae sound system has been set up in the park in front of the community centre. Children dance in an amphitheatre strewn with bean-bags. Smells and noise emanate from the diesel generators lighting and powering the stalls. Gözleme, a Turkish flatbread that is a staple at markets across Sydney, is popular, but no one is buying from the gelato cart tonight. The mix is hipster meets community group as pulled pork sandwiches and badges with the red, yellow and black of the Aboriginal flag change hands.

Outside the front door of the community centre are two representatives from Urban Growth NSW, who in their literature describe themselves as 'The NSW Government's city transformation arm'. Their set up includes a table with some pamphlets, and two blackboards behind them on which is scrawled 'Great creative placemaking ideas and inspiration here'. They are here for 'community outreach' on the 'Central to Eveleigh

Urban Transformation and Transport Program', which the brochures say is 'about great city thinking that connects diverse and vibrant communities'. Redfern is not mentioned in this project, except for a single reference to Redfern Railway Station, even though it will undoubtedly impact on the local community. Perhaps the insistence on the name Eveleigh is due to it being *not*-Redfern.

The movement between local and global registers at the Redfern Night Markets is congruent with the local council's global brand 'City of Villages'. In support of its brand, the City of Sydney council commissioned a free booklet, which in its aesthetic referenced Moleskine notebooks and vintage Penguin Classic book covers. Inside, Redfern was described as

> [combining its] Aboriginal roots and working class history with modern tribalism. The unvarnished, industrial atmosphere of these neighbouring pockets has lured contemporary art circles, vintage/retro furniture restorers and slow food aficionados. The nostalgic flavour of this village is deep with respect for what has gone before... Redfern... people are active lovers of change and reflect Australia's multi-ethnic population. They operate with a DIY mentality and loyalty to the local economy, the original landowners the Cadigal people and you, if your ear is open.
>
> (City of Sydney, 2011b, pp. 41–43)

Certainly, the Redfern Night Markets work as a facilitator between the Aboriginal community and the creative economy, inviting visitors into the Redfern Community Centre. The community centre opened in 2004 in a converted shoe factory, after a lengthy campaign against it from local non-Aboriginal residents on the basis that services and spaces that cater specifically to Aboriginal people would further entrench The Block as a place for Aboriginal people to meet up and to stay in Sydney (Shaw, 2007). Across the road from the centre and the night markets, the Redfern Aboriginal Tent Embassy occupying The Block is a reminder that the type of visibility offered to Aboriginal culture by the Creative City does not translate into other types of representation for Aboriginal people, particularly with regard to urban planning and governance. As Behrendt notes:

> One of the consequences of overlooking Indigenous presence and experience is to exclude us from effective participation in civic life, notably social policymaking – whether in areas specifically relating to Aboriginal people themselves or broader collective decisions in areas such as town planning and urban development.
>
> (Behrendt, 2006, p. 7)

To varying degrees, both Redfern Night Markets and Carriageworks Farmers Market are indexical to the problematics of urbanism as it is being produced in globalized city centres, where community, place and making are being mediated by discourses of urban renewal, gentrification, and the Creative City. This urbanism purports to be about making place, but is too often based on displacement. In Redfern, this discursive urbanism is further problema- tized by its articulation through whiteness and the postcolonial spatial, social and cultural legacies of *terra nullius* (Shaw, 2007). In the concluding chapter of the book, I look at the urbanism of markets again, not as a form of place- making in cities, but as a way of making place in the city that is everyday and sustainable. First, however, in Chapter 6, I want to look at another farmers' market in San Francisco, one that makes place and provides valuable mater- ial infrastructure for marginalized communities, and in doing so, actualizes a right to the city.

6   The right to the city in a downtown farmers' market

San Francisco

The Heart of the City Farmers' Market sets up twice weekly in front of San Francisco's City Hall in UN Plaza. The plaza is an official public monument to the 1948 Universal Declaration of Human Rights, and is located in Civic Center, the seat of political power in San Francisco. UN Plaza also borders Tenderloin, one of the city's most deprived neighbourhoods. The market sells olive oil, sunflowers, organic kale and other products that signify the rural in a space that is inescapably urban. Food trucks, which are synonymous with hipster foodie-ism sell wood-fired pizza and vegan snacks in close proximity to a statue of Simón Bolívar, a figure of revolutionary struggle and liberation politics. Homeless men and women sit in the sunshine next to workers grabbing a bite to eat in their lunch break. There are stalls stocking everyday staples for Mexican, Chinese and South-East Asian cooking. An informal market of older people buying and selling groceries out of plastic bags meets on the fringes, and is packed up and gone by the time I walk to the end of the market and back. Heart of the City is a place where rural and urban, poverty and affluence, the everyday and cultural capital, informal and formal, disadvantage and power meet. Here, the palpable dialectics of this market, and an associated 'rubbing along' with difference (Watson, 2006, 2009), make place. Previously, I have suggested that the heterogeneity of markets is metonymical to the heterogeneity of the city. In this chapter, I want to suggest that this metonymy, as played out between Heart of the City and the city of San Francisco, is an actualization of Henri Lefebvre's (1996) call for the right to the city.

Making as the right to the city

'The *right to the city* cannot be conceived of as a simple visiting right or as a return to traditional cities. It can only be formulated as a transformed and renewed *right to urban life*,' writes Lefebvre (1996, p. 158, emphasis in original). By this he means that

> The right to the city, complemented by the right to difference and the right to information, should modify, concretize and make more practical the rights of the citizen as an urban citizen (*citadin*) and user of multiple services. It would affirm, on the one hand, the right of users to make

known their ideas on the space and time of their activities in the urban area; it would also cover the right to the use of the center, a privileged place, instead of being dispersed and stuck in ghettos (for workers, immigrants, the 'marginal' and even for the 'privileged').

(Lefebvre, 1996, p. 34)

Lefebvre's call for a right to the city for *citadins* has been influential in recent urban theory and practice. His work has increasingly been interpreted as a spatial or material right to the city, and the search for spatial analogues of the right to the city frequently turns up equivalences such as 'public access to urban space' (Merrifield, 2014). However, Andy Merrifield clarifies that the right to the city as Lefebvre formulates it should not be understood to be 'a tourist trip down memory lane, gawking at a gentrified old town; nor is it enjoying for the day a city from which you've been displaced' (2013, p. 24). The right to the city is not a measure of access to urban space, 'but to urban life, to renewed centrality, to places of encounter and exchange, to life rhythms and time uses, enabling the full and complete *usage* of these moments and places, etc.' (Lefebvre, 1996, p. 179, emphasis in original). It is the right to *be* urban, where the urban is a phenomenon that emerges from making that is playful, creative and sensorially complex. Being urban in the Lefebvrian sense is being mixophilic, experiencing pleasure and feeling at ease with difference. Lefebvre expounds upon this:

The human being has the need to accumulate energies and to spend them, even waste them in play. He [*sic*] has a need to see, to hear, to touch, to taste, and the need to gather these perceptions in a 'world'. To these anthropological needs which are socially elaborated (that is, sometimes separated, sometimes joined together, here compressed and there hypotrophied), can be added specific needs which are not satisfied by those commercial and cultural infrastructures which are somewhat parsimoniously taken into account by planners.

(Lefebvre, 1996, p. 147)

It is easy to see, reading Lefebvre's words, why the right to the city has become a rallying cry for artists, creative practitioners and creative workers (Begg and Stickells, 2011) – those who are quite literally 'creative' in the city – and a vehicular idea (McClennan, 2004; Peck, 2012) for modes of urbanism (everyday, guerrilla, insurgent, DIY, tactical, user-generated) that deploy creativity and the Creative City as modes of intervention (Iveson, 2013; Bela, 2014).

The connections with creative practice are strengthened by Lefebvre's assertion that the right to the city is the participation in making that is creative, ludic, gratuitous, sensorially led and, ultimately, because of these attributes, urban. When Lefebvre talks of making it is not of making as generative correspondence and emergence as Ingold (2011) does. In Lefebvre, making is shaping things: 'Indeed, if they have influenced urban rhythms and spaces, it is be enabling groups to insert themselves, to take charge of them, to *appropriate* them; and this, my inventing, by sculpting space (to use a metaphor), by giving themselves rhythms' (1996, p. 105, emphasis in original). He refers to the plasticity of space and its responsiveness to modelling (1996, p. 79). Harvey picks up on this in his reading of Lefebvre, referring to the right to the city as a form of making:

> it is a right to change ourselves by changing the city. It is, moreover, a common rather than an individual right since this transformation inevitably depends upon the exercise of a collective power to reshape the processes of urbanization. The freedom to make and remake our cities and ourselves is, I want to argue, one of the most precious yet most neglected of our human rights.
>
> (Harvey, 2008, p. 23)

Harvey's language also suggests a model of making based on material and structural processes. The shape that emerges from making that is *urban* is the city itself.

The city as the spatial realization and performance of participatory making that is urban in disposition is an instance of what Lefebvre calls the *oeuvre* (work). Significantly, given the subject of this book, Lefebvre's other exemplar of the *oeuvre* is a close manifestation of the market: the festival (*la fête*). Lefebvre writes, 'the eminent use of the city, that is, of its streets and squares, buildings and monuments, is *la fête* (which consumes unproductively, without any other advantage than pleasure and prestige)' (1996, p. 66). If we follow Lefebvre's thinking, then markets become truly urban when, as I wrote in Chapter 1, the product or exchange function was separated out to become 'the market' (Agnew, 1986). Subsequently, the use value or *oeuvre* of the marketplace came to the fore. Markets, like the city and the festival, are a 'place of encounter, the assemblage of differences and priority of use over exchange value' which is produced from making that is, in Lefebvrian terms, urban. Markowitz (2010) in assessing the value of farmers' markets to low-income consumers noted that the value was not purely related to the provision of fresh food, but was in the act of participatory making as well. I would suggest, then, that markets like Heart of the City, which are everyday, collective, convivial, heterogeneous, inclusive and liminal, and place that emerges from making in these types of markets, are instances of Lefebvrian *oeuvres*, and that the right to the city is also the right to make urban place.

'For Lefebvre, it was not the home, but the city, which expressed and symbolized a person's being and consciousness' write Kofman and Lebas (1996, pp. 7–8). This type of individual and collective access to the urban is not possible in the private domestic sphere of the home, nor is it as possible in the suburbs or the small town. Lefebvre's insistence on this is not motivated by a city-centric snobbery, but a realization that the conviviality and sociality of the urban cannot exist in spaces that do not encourage collective, public, spontaneous encounters with difference. As Merrifield explains, for

Figure 6.1   Heart of the City Farmers' Market, San Francisco

Lefebvre, '[t]here can be no city without centrality, no urbanity,... without a dynamic core, without a vibrant, open public forum, full of lived moments and "enchanting" encounters' (Merrifield, 2013, p. 12). Writing about San Francisco, Solnit says that creeping suburbanization is the death of the urban because it represents a turn away from the public sphere:

> The sense of the city as home is eroding as the public sphere ceases to be a place where people feel at home;... the new arrivals seem to live in it as though it were a suburb... Urban life, like cultural life, requires a

certain leisure, a certain relaxation, a certain willingness to engage with the unknown and unpredictable. For those who feel impelled to accelerate, the unknown and the unpredictable are interference as the city's public space becomes not a place to *be* but a place to traverse as rapidly as possible.

(Solnit and Schwartzenberg, 2002, p. 123)

However, the city as the *not*-home does not disallow the possibility of feeling *at home* in and with the heterogeneity of the city or, for that matter, of markets. This is, as Kofman and Lebas have noted, increasingly difficult in globalized cities where 'The logic of the market has reduced these urban qualities to exchange and suppressed the city as an *oeuvre*' (1996, p. 19). As evidence of this, Merrifield points to the 'recent phenomenon' where 'cities themselves have become exchange values, lucre *in situ*, jostling with other exchange values (cities) nearby, competing with their neighbors to hustle some action – a new office tower here, a new mall there, rich flâneurs downtown, affluent residents uptown' (2013, p. 23). Similarly, the globalizing processes that have affected local place and markets (the impacts of which I have discussed at length in this book) could surely be viewed as suppressing capacities as oeuvre.

A downtown farmers' market

When we think of making and places that are urban in a Lefebvrian sense, and which therefore actualize a right to the city, farmers' markets do not come readily to mind. In the popular imagination, they are a bourgeois fancy, an outlet for affluent consumers who can afford to pay a premium to consume ethically. In a book of drawings and stories of everyday life in San Francisco titled *Meanwhile in San Francisco: the city in its own words*, local illustrator Wendy MacNaughton sketches a hipster from The Mission – a neighbourhood that is a metonym for the vertigo-inducing city-wide gentrification of San Francisco – with a bag in-hand full of farmers' market produce (2014, p. 93). This portrait confirms, and conforms to, a perception of the farmers' market as a space and a set of practices that are, above all, 'middle-class' and that deploy cultural competence as a mode of distinction (Bourdieu, 1984). This perception is not solely drawn from the collective imaginary. Markowitz points out that 'Although the enactment of community occurs on weekend mornings in nearly every American city, inequalities of class and geography preclude many' (2010, p. 66). Moreover, the links between farmers' markets and middle-class taste have been investigated within the context of whiteness (Alkon, 2012; Guthman, 2011; Alkon and McCullen, 2011), ethical consumption (Alkon, 2012; Lewis and Potter, 2011) and gentrification (Zukin, 2008).

Lefebvre is adamant that the right to the city is the right to being, to making, and to place that is urban, not rural. Farmers' markets in the city, on the other hand, amplify representations of rural place and practice. The semiotics of the farm as healthy living and the source of fresh, high-quality produce have persisted despite widespread documentation of the costs of industrial-scale agricultural production including environmental destruction through unsustainable land use and farming practices; the exploitation of labour; public health concerns around pesticides and food safety; and animal welfare abuses. Imagery of small-scale farms and farming, and cheerful farmers are used to advertise even those nodes in the food industry where locally produced food is anathema – from fast food restaurants to pre-prepared and processed food products, and supermarket chains. These images are products of the 'agrarian imaginary' (Mayes, 2014), an urban phenomenon that nostalgically perceives agriculture as unalienated labour and a pre-industrial, artisanal mode of production. First-hand accounts from labourers at Heart of the City dispute the romanticized view of agricultural labour. When MacNaughton interviews one of the workers at the market for her book on San Francisco, the worker tells her: 'Our family came from Vietnam. We were the first Asian family in Rio Linda. Do you know about Rio Linda? There were lots of KKK... We get up at 2am, drive many hours, and set up everything for the early customers... We work hard... It used to be tough here' (2014, pp. 44–63). The farmer was unequivocal about the realities of their job – that it was back-breaking work for little financial return. Theirs is a family-run business, in which both parents and nine out of 11 children work. All the children have other jobs in addition to their market duties.

Farmers' markets, in spite of the associations with exclusivity, are following global trends and becoming more common in the US. Voluntary, self-reported data collected by the United States Department of Agriculture (USDA) shows that their numbers have been increasing every year since the tabulation was initiated in 1994. In 2014 there were 8,268 farmers' markets in the US (National Count of Farmers Market Directory Listing Graph: 1994–2014, n.d.). Heart of the City is a certified farmers' market (CFM). In San Francisco, certification is administered by the California Department of Food and Agriculture (CDFA), but overseen by the San Francisco Department of Public Health's Agriculture Program. The California Code of Regulations stipulates that a CFM is

> a location approved by the county agricultural commissioner of that county where agricultural products are sold by producers or certified producers directly to consumers or to individuals, organizations, or entities that subsequently sell or distribute the products directly to end users. A certified farmers' market may only be operated by one or more certified producers, by a non-profit organization, or by a local government agency.
>
> (California Code of Regulations, 3 CCR § 1392.2)

CFMs have existed in California since 1977, when regulations were relaxed regarding the standardization of sizing, labelling, packaging and transportation of produce for external sale. The rationale for decreasing regulation was to

> provide a flexible marketing alternative without disrupting other produce marketing systems. The high quality and fresh produce brought to the CFM's by its' producers creates a diverse market and also provides the consumer with opportunity to meet the farmer and learn how their food supply is produced. CFM's provide a great opportunity for small farmers to market their products without the added expenses of commercial preparation [*sic*].
>
> (Certified Farmers Markets Program, n.d.)

The first farmers' markets to open in California after the easing of packaging and distribution requirements were started by 'food system advocates' whose primary concern was food justice (Project for Public Spaces, 2003, p. 12) Farmers' markets in California are a platform for delivering federal, state and local public health, community development and education programmes aimed at mitigating the effects of diet-related disease through the provision of affordable, fresh food and the promotion of healthy eating in low-income areas and in food deserts (Young et al., 2011). According to the USDA, food deserts are 'urban neighborhoods and rural towns without ready access to fresh, healthy, and affordable food… and are served only by fast food restaurants and convenience stores that offer few healthy, affordable food options' (Food deserts, n.d.). As well as having spatial indicators, food deserts have social indicators and are '"low-income communities", based on having: a) a poverty rate of 20 percent or greater, OR b) a median family income at or below 80 percent of the area median family income' (Food deserts, n.d.).

Heart of the City's website says that the adjacent area of the Tenderloin is 'a low-income food desert with a high rate of preventable, diet-related disease and an average life expectancy 20 years lower than surrounding neighborhoods' (About our market, n.d.). The San Francisco Department of Public Health agrees that access to fresh produce is an issue for residents of the Tenderloin. Susana Hennessey Lavery (2014), a health educator with the department, points out that the challenge is not that there is no food for sale at all. There are more than 70 corner stores serving the Tenderloin. Rather, there is no full-service supermarket or grocery store in an area that, according to US Census data, had a population of 31,547 people in 2010. Lavery prefers the term 'food swamp' (2014, n.p.), which is materially correct in its implication that food is available, whereas 'food desert' as a term implies an absence of food supplies. The latter term is more commonly used by government organizations (Food deserts, n.d.), non-profits and academic researchers in applied urban contexts to delineate an area that is underserved, and where food that is not nutritionally beneficial is still available. Both terms are value-laden. 'Desert' suggests a landscape barren of food, suggesting that any food available is not worthy of being categorised as food, and therefore not fit for consumption; 'swamp' connotes a murky terrain, impure, unclean and, again, something not fit for consumption.

Definitions and values about what constitutes 'food' aside, Heart of the City, unlike the farmers' market in Sydney that I wrote about in the previous chapter, was not set up as a vehicle for urban renewal. Heart of the City was established in 1981 as an explicit counterpoint to the degradation of place in San Francisco's downtown area caused by economic and social disadvantage, homelessness and failing infrastructure. In this market, food justice is imbricated with spatial justice. The market's mission is to provide material and social infrastructure for a community living in a food desert (About our market, n.d.). The market accepts EBT benefits (food stamps), engages in advocacy and runs a number of health and education initiatives aimed at increasing access to fresh food and supporting healthy eating in the local community (About our market, n.d.). Heart of the City's objectives, and the civic-bureaucratic framework in which it functions are at odds with popular and collective perceptions of farmers' markets as a high-end marketing category and as the site for the performance of middle-class consumer practices. To the contrary, empirical evidence has demonstrated that prices in farmers' markets in California are no more expensive than at small grocery stores or supermarkets, and deliver higher returns to farmers (Young et al., 2011, p. 211). Instead, Heart of the City is a place where the right to the city exists for the groups for whom it is most pressing and meaningful: the socially and spatially disadvantaged (Merrifield, 2013). Kofman and Lebas remind us 'that for Lefebvre rights are not simply derived from the politico-State level but are also anchored in civil society' (1996, p. 41). Heart of the City is an example of how these rights can be achieved in practice.

The heart of San Francisco

The role that Heart of the City has taken up in advocating for food justice in the Tenderloin is not inconsistent with the broader political agendas that farmers' markets perform around ethical and sustainable consumption. Farmers and farm workers, too, are at the forefront of global social movements, such as food sovereignty (Trauger, 2015), slow food (Petrini, 2003, 2007), land reform (Wolford, 2004) and anti-globalization (Bové and Dufour, 2001, 2005), which position food justice (Gottlieb and Joshi, 2010) as inseparable from social and environmental justice. Alison Alkon (2012) situates farmers' markets in the Bay Area within a genealogy of local radical politics that includes the Black Panther Party (founded in Oakland in 1966) and student activism at UC Berkeley in the 1960s.

San Francisco as a city of radical, community-minded politics is an objective and subjective place-myth that has persisted among San Franciscans and for the global community. Local writer Rebecca Solnit says that San Francisco 'used to be the great anomaly. What happened here was interesting precisely because it was different from what was happening anywhere else.

We were a sanctuary for the queer, the eccentric, the creative, the radical, the political and economic refugees, and so they came and reinforced the city's difference' (Solnit and Schwartzenberg, 2002, p. 31). Her essay 'The city's tangled heart' recounts how Civic Center, where Heart of the City stands, is the site for many critical moments in the formation of place in San Francisco. Solnit's cartography marks out points of psycho-geographic energy: the spot where racist riots broke out against the city's Chinese community in 1904; the *TRUTH* mural by local artist Rigo23, that overlooks UN Plaza and is dedicated to the Angola Three, a trio of Black Panther activists imprisoned in solitary confinement for decades; and City Hall itself, where gay rights activist and politician Harvey Milk was shot in 1978, and where, 16 years later, the first same-sex marriages in the United States took place. Solnit writes:

> United Nations Plaza, poised between San Francisco's Civic Center and skid row, is also the axis along which stand many monuments, both obvious and unknown, to the history of race, justice, and power, a version of the heart of the city, this map tries to make visible ... This map attempts to recognize only some of the turbulence that swirled around San Francisco's administrative heart, which is also a heart of struggle, of suffering, and sometimes of overcoming.
>
> (Solnit et al., 2010, p. 37)

Kathleen Bubinas advances that 'identifiers and place names' for farmers' markets 'resonate with the local population as a marketplace that is distinctively embedded in the community and realized through a shared history, ethnic heritage, class affinity, and value system' (Bubinas, 2011, p. 156). Heart of the City is an evocative name for a market. Its semiotics are impossible to ignore. The heart as a trope connotes love, compassion and generosity. It evokes the image of an essential urban core, beating rhythmically, reliably as it sends life-giving blood and oxygen to the city's extremities. The 'heart of the city' is also a spatial reference denoting the location of the market at the centre of San Francisco's urban core in Civic Center, the ostensible heart of the city's political system.

As you walk through the market, excerpts from the 'Preamble' to the *Charter of the United Nations*, underfoot in the plaza paving, assert the universality of human rights: 'to reaffirm faith in fundamental human rights, in the dignity and worth of the human person, in the equal rights of men and women and of nations large and small' (see Figure 6.2). Heart of the City, surrounded by poverty and wealth, offers a respite, or at the very least a neutral place, in a city that is divided by uneven urban development. Solnit has in recent years been writing about the threats to and conflicts over place posed by large-scale urban renewal, gentrification and privatization of public space and infrastructure within the context of the ascension of Silicon Valley. The denigrations of the public sphere in a city that has prided itself on tolerance for alternative and outsider politics and ways of living display an increasing

Figure 6.2   UN Plaza, Civic Center, San Francisco

mixophobia. Solnit recounts how the place-image of San Francisco as a city with permissive attitudes towards individualism is derived from the home it has provided throughout its history to the outlier, the counter-cultural, the alternative, the idiosyncratic, the eccentric. However, in contemporary San Francisco, that individualism has been recast as the infinite possibilities for the right to the city for the (white, male) creative tech entrepreneur (Packer, 2013). Solnit wrote of the dot.com boom that San Francisco experienced in the late 1990s, 'Whatever the Internet may be bringing the masses stranded far from civilization, the Internet economy in its capital is producing a massive cultural die-off, not a flowering' (Solnit and Schwartzenberg, 2002, p.31).

Solnit characterizes the struggles over place in the past two decades as a battle for the heart of the city between old San Franciscos (counter-cultural, alternative, inclusive) and new San Franciscos (disembodied, tech, exclusive), but also between types of gentrification (Tonkiss, 2005). For Solnit, previous taste-based gentrifiers valued 'social diversity; it is what makes the city a thriving ecosystem, a garden in which the weeds sometimes bloom most brilliantly' (Solnit and Schwartzenberg, 2002, p. 78). Current gentrifiers are 'monoculture. The expensive new businesses coming to the Mission include a plethora of restaurants and houseware and clothing stores, but bookstores, theatres, dance studios, galleries, "monoplex" movie houses and non-profit activist organizations are shrinking, not multiplying, with this stage of gentrification' (2002, p.78). Lefebvre insists that improvised encounters with difference and diversity are one of the defining qualities of urban place, and indexical to a right to the city for all *citadins*. Monoculture obstructs the right to the city because it suffocates the urban. Heart of the City Farmers' Market is the heart, and is at the heart of a city that is disappearing: 'What is happening here eats out the heart of the city from the inside: the infrastructure is for the most part being added to rather than torn down, but the life within it is being drained away, a siphoning off of diversity, cultural life, memory, complexity. What remains... will be a hollow city' (Solnit and Schwartzenberg, 2002, p. 30).

Away from the heart of the city

Heart of the City is a material, visible and centrally located (re)affirmation of the social value of public space in a city where the 'broad decay of twentieth-century American liberalism' (Low and Smith, 2005) has had direct consequences on the making of place from urban space. A more common scenario was evident in another market that we visited in the San Francisco Bay Area, the Alameda Point Antiques Faire. In 1974, anthropologist Robert Maisel published an ethnography of the Alameda Penny Fair, the Faire's precursor-of-sorts. Maisel's account of the Penny Fair substantiates a place-image of the Bay Area as the home of alternative culture and politics. Maisel discerned a mixophilia or, at the very least, a tolerance towards social and cultural difference at the market. Maisel concluded that place and atmosphere at the Penny Fair emerged from this mix:

> The seeming over-representation of counter-culture youth (thirteen percent of the vendors were so identified by their appearance) is not unexpected in this locality...
>
> The easy atmosphere (or 'good vibes') which many vendors find attractive is due to the lack of 'hassle' they encounter... Vendors are rarely chastised for doing their thing, whether it be nursing babies, smoking grass, setting up mini-encampments, drinking beer or appearing in various sorts of casual dress (and undress). Perhaps flea marketers feel unconscious pride in being participants in a social activity where blacks and 'red necks,' hippies and squares, homosexuals and straights visibly rub shoulders as they share the common foci and concerns of the market.
>
> (Maisel, 1974, p. 493)

The Penny Fair closed in 1995. In 1998, the Alameda Point Antiques Faire, a monthly meet-up of second-hand dealers and buyers, opened. When we visit the latter, four decades after Maisel went to the Penny Fair, any trace of the urban – be it spatial, social, cultural – is hard to detect. Few of the elements identified by Young et al. (2011) as conducive to a successful market were at play at Alameda. The spatial design of the Faire goes against research that suggests that the factors for a good location for a market include 'walkability, visibility, parking… [and] the ability of a location to draw customers to or from other activities, such as shopping, churchgoing, and recreation' (Young et al., 2011, p. 211).

The Antiques Faire does have the hallmarks of a second-hand market in that it is located on the urban periphery, on what used to be the runway of a United States Navy airbase at the end of Alameda Island. We arrive by public transport on the ferry from San Francisco. The distance from the ferry terminal may be short, around one and a half kilometres, but the terrain is not pedestrian-friendly. Most visitors are in the long, steady, slow-moving queue of cars that accompanies us as we tramp through the mud along the side of the road. On one side, we are flanked by the Bay and the giant cranes and stacks of shipping containers at the Port of Oakland. On our other side are acres of fenced-off cracked tarmac and pampas grass. The road ends in a vast car park that makes maximum use of the decommissioned airbase. A large proportion of online reviews of the market comment on or offer advice on negotiating the parking lot (Alameda Point Antiques Faire, n.d.). Sifting through these reviews, I retrospectively learnt that there is a shuttle bus to navigate the parking lot, but we duck and weave the cars on foot until we reach another threshold: a ticket booth. Entry to this market requires lining up and purchasing a ticket. This is not the porous, public market that we are used to. Once inside the perimeter fence, the market is huge, both in the scale of the site and the scale of consumption. There are shopping trolleys for rent, and several people push industrial sized carts loaded up with mid-century furniture and distressed gas station signs. There is a drive-in section at the side of the market where large and bulk purchases can be loaded straight into your car.

The size of the space inhibits the sociality of the market. The stalls are far apart from each other. The in-between spaces where buyers walk are expansive. They need to be to enable the free movement of the trolleys of furniture. There is no bumping into each other or making room for other people. Camaraderie with the next-door stall is not easy when there is no proximity. There were faint traces of the Bay Area's signature irreverence for mainstream politics. Someone had attached a hand-drawn speech bubble saying 'Don't vote for my son. You'll be sorry…' on a vintage George Romney 'Romney for Leadership' campaign banner. (This was the week before the Obama–Romney presidential election.) However, in terms of the focus of the market, it would seem that the commerce and entrepreneurialism that Maisel (1974) isolated as the field of action at the Alameda Penny Fair is still very much the field of action at the Antiques Faire, and the social aspect of markets, which Maisel underplayed as a 'myth', was also not a definitive quality here. As a place, the atmosphere did not invite you to linger. There were no trees or shady spots to shelter out of the sun. When I looked for somewhere to sit down and feed my son, the few seats available were all occupied and right next to some overflowing garbage bins. There were also some saggy couches in an unappealing 'customer lounge' in a demountable trailer near the entrance. The food and drink options were limited to traditional fast food, except for one food truck that sold fusion *bao* and a stall that sold artisanal iceblocks (a poetic, although perhaps unintentional link to Alameda's history as the birthplace of the popsicle).

The centrality and spatial accessibility of Heart of the City is a point of difference from the Antiques Faire. The latter is not located close to any other social or commercial infrastructure, except for the storage facilities and a distillery that have set up in the disused airbase buildings. In contrast, Heart of the City encourages Lefebvre's idea of the urban in a landscape where planning and infrastructure have not always been ambitious in encouraging the types of making that produce urban place. Even UN Plaza, in spite of its visual rhetoric about the value of human rights, is not designed at the human scale and is usually empty and underutilized, and hence not particularly welcoming at the times when Heart of the City is not there.

The market worker whose story is narrated by MacNaughton talks of Heart of the City as if it is home, acknowledging that the market has value beyond the surplus value of their labour: 'we take care of each other. Give IOUs. Hire the homeless to help load and unload. Help the older people. It's like a family. I started coming here thirty years ago this June. I said I'd stay a year' (MacNaughton, 2014, pp. 58–63). Heart of the City challenges neo-liberal political and media discourses and urban spatializations that degrade the value of public space, and revalorizes it through making place. Through encounters with heterogeneity and participatory making that are urban, San Francisco's *citadins* simultaneously produce the *oeuvres* of market, place and city. If, as Bubinas claims, 'each [farmers' market] takes on a persona that is unique to place and people – the organisers, vendors, artisans, musicians, and customers who comprise it' (2011, p. 156), then the persona of the Heart of the City Farmer's Market is urban as defined by Lefebvre, and as such, it realizes a right to the city.

7   Markets and urbanism

Antwerp

In this final chapter, I look at a particular ecology of market, place and city in order to explore the connections between markets and urbanism. Every Sunday Antwerp's Vogelenmarkt (sometimes known as Vogelmarkt or Vogeltjesmarkt) makes place in Theaterplein, a public square in the city's historic centre. The urbanism of Vogelenmarkt and Theaterplein is a hybrid urbanism. It is the pre-modern urbanism of the marketplace and the public square (Calabi, 2004; Tangires, 2008; Zucker, 1970). At the same time, the assemblage of market and public space references the urbanism of the modernist city through the monumentality and order of Bernardo Secchi and Paola Viganò's 2009 redesign of the square. Secchi and Viganò's attention to the centrality of place in their design also references contemporary urbanisms that are place-based and concerned with the human scale (Gehl, 2010; Lydon and Garcia, 2015). Finally, I want to propose that Vogelenmarkt and Theaterplein, together, produce a future-oriented urbanism that is everyday and sustainable.

Vogelenmarkt + Theaterplein

Theaterplein provides the stage for two of Antwerp's weekly markets. In addition to Vogelenmarkt on Sundays, there is a Saturday market colloquially known as Exotische Markt (Exotic Market) because of its trade in street food, spices and imported goods. As its epithet suggests, the Saturday market is a performative display of multiculturalism in the city's centre. It is an exhibition of cultural diversity – as distinct from the habitus of cultural difference (Bhabha, 1994) – that is generated, disseminated and thus ultimately controlled from within the dominant cultural paradigm. 'Exotic', as a descriptor or category, is discerned only when positioning something as other. Ghassan Hage, in talking about discourses of multiculturalism in Australia that have been generated within dominant Anglo-Australian culture, has said that cultural diversity is often articulated through food. As an example, he cites the 'multicultural fair where the various stalls of neatly positioned migrant cultures are exhibited and where the real Australians, bearers of the White nation and positioned in the central role of the touring subjects, walk around and enrich themselves' (1998, p. 118). At Exotische Markt, as in the multicultural fair, any unfamiliarity or unease with cultural difference is smoothed out and over. The Saturday market is not commonplace diversity (Wessendorf, 2014) in action, even though Antwerp is socially and culturally super-diverse (Blommaert, 2013). Vogelenmarkt, in comparison, is not a remarkable market. Its site in the centre of the city mediates the goods and clientele, but it is, nevertheless, an everyday market with an unremarkable mix of stalls selling inexpensive and ubiquitous goods such as fresh food, fabrics, household cleaning products and mass-produced clothing. Up until 2015, when the city of Antwerp introduced trading hours on the first Sunday of the month (Sunday Shopping, n.d.), the most remarkable aspect of Vogelenmarkt was that it was one of the only options for stocking up on necessities on a Sunday.

Vogelenmarkt belongs to the market culture of a city whose history is spatially and culturally entangled with mercantilism (Calabi, 2004; Mumford, 1991 [1961]). The place-myth of commerce has proved to be so enduring that it forms part of Antwerp's place marketing. The city's official tourism portal promotes 'Maritime' and 'Diamond' themed itineraries for visitors (Visit Antwerpen, n.d.), drawing on Antwerp's history and identity as an international port and its central position in the international diamond trade. In keeping with the dominance of mercantilism in the city's history, many of Antwerp's major shopping streets and squares are named for the markets that used to take place there; for example, Eiermarkt (Egg-market) and Schoenmarkt (Shoe-market). Antwerp's landmark main square, Grote Markt (Great Market) is a major tourist destination and is still the stage for market-like events and fairs throughout the year. Grote Markt is overlooked by the city's town hall, which is in line with urban planning in medieval and early modern Europe (Simms, 2014) when marketplaces were situated within the purview of local authorities so as to monitor the collection of taxes, the quality of goods and the conduct of buyers and sellers. Another square in Antwerp's old centre, Vrijdagmarkt (Friday Market) has, more or less, continuously hosted a second-hand market and auction every Friday since the sixteenth century.

In Vogelenmarkt, which has also existed since the sixteenth century, the birds in question were originally poultry and game birds, though a trade in imported birds certainly resulted from Antwerp's central role in Belgian colonialism. The British artist Mark Dion commented on this lugubrious trade in his 1997 installation *The birds of Antwerp*, which incorporated live finches that were purchased at the Vogelenmarkt (Bryson, 1997). Nowadays, the bird section (which has been expanded to include other small domestic animals such as rabbits and guinea pigs) is the most spectacularized section of the market, judging from its visual dominance on tourism portals and social media.

Vogelenmarkt is remarkable because it is an unremarkable market in the gentrified, touristic centre of a twenty-first-century city. On the other hand, Theaterplein, laid out in front of Antwerp's Stadsschouwburg (Municipal

Theatre) and Het Paleis, a children's cultural centre, has been redesigned to be a remarkable public square – which is hardly remarkable within the spectrum of public spaces in globalized cities (see Miles, 2010). The award-winning design by the Italian architects and urbanists Bernardo Secchi and Paola Viganò created a multi-purpose canopy of cathedral-like proportions to disguise the theatre's Brutalist façade. On Sundays, when the market is set up, the monumentality of the canopy is incorporated into the market infrastructure and provides protection from the elements for stallholders and the market crowd. Its success is that you hardly notice it. It disappears. In spite of its visual rhetoric, which is modernist rather than vernacular, Secchi and Viganò's design adapts to the 'human scale' advocated by contemporary urbanists (Gehl, 2010; Oldenburg, 1999) through its housing of the market. In spite of the canopy's epic dimensions, I did not notice it at all when I first visited Vogelenmarkt, and only become aware of it when I returned later in the week when the market was no longer there.

Theaterplein was not conceived as a single intervention in urban space, but one element of *The structure plan for Antwerp* (2003–2006), a city-wide plan commissioned from Secchi and Viganò by the city of Antwerp (Secchi and Viganò, 2009). Their urban plan shifts between macro- and micro-perspectives of Antwerp, and produces manifold coexisting urbanisms: pre-modern and modernist; monumental and people-centred; centralized and improvisational. Secchi and Viganò's designs do not fixate on space, like the urbanism of modernity and modernism, nor do they obsess over place as contemporary urbanist theory and practice often does. Secchi and Viganò emphasize heterogeneity, or what they call 'mixité' (2009, pp. 137–139), as an objective in the redesign of Theaterplein, both in terms of uses and users. They were also keen to respect the qualities of place that Vogelenmarkt already organically produced. Unlike the redevelopments of many other market spaces under the guise of urbanism (and its frequent stand-ins gentrification and urban renewal; see Smith, 2002), Secchi and Viganò's redesign of the public *space* of Theaterplein for the *place* of Vogelenmarkt did not entail transforming the remit of the market or the community it was serving. Instead, Theaterplein and Vogelenmarkt produce an image of the city as organism 'where the diversity of the city [is] a system of specialized organs functioning for the common good of the corporate body. Each part of the body has a role to perform, and this performance benefits the well-being of the entire body' (Langer, 1984, p. 101).

Urbanism and place

When the market is on, Theaterplein is crowded and lively from the movement of people and materials. The market's spatial and social dimensions are human-scaled and characteristic of a place-based urbanism that depends on engagement with local communities and local (built) environments (Haas, 2012). The urbanism of the square on market day is a vestige of the pre-modern city because a market is

Figure 7.1 Market day: Vogelenmarkt, Theaterplein, Antwerp

the physical expression of an oral, pedestrian society, in which growers, makers and consumers often faced each other directly. It sells products of all kinds, old and new; raw, hand-made and manufactured; and usually unpackaged and unbranded. The price is established by negotiation or haggling.

(Davison, 2006, p. 4)

Parham (2015) outlines how markets have adapted with morphological, technological and social developments, yet the structure of local place-based markets

Figure 7.2   Theaterplein, Antwerp

has not changed radically since pre-industrialization (Geertz, 1963; Visconti et al., 2014). In spite of their historical entanglement with cities and capitalism, markets are, in their most basic form, pre-urban and pre-capitalist. In a global political economy where new iterations of the market dominate they are vestigial. This out-of-time, heterotopic (Foucault, 1973; Foucault and Miskowiec, 1986) quality of markets explains their popularity and nostalgic appeal for groups such as maker and crafting cultures who seek alternative sites and modes of production, distribution and consumption (Sully and Westbury, 2015).

Place and atmosphere change in Theaterplein when the market packs up. A hyper-efficient team of cleaners moves in, washing away any residue of a market with their industrial-strength hoses. All that remains as material evidence is the permanent power points that the market stalls plug into for electricity. Without the eye-level dimension of the market, Theaterplein's space discursively shifts from local and prosaic to bureaucratic and mythological, and this is reflected, too, in the accompanying temporal and spatial transformations of place. When it reverts to monumental gateway to the theatre on weekdays, the architecture assumes prominence over the human. The making that is happening here changes. People make lines across the square as they walk the dog, or from A to B. There is a skateboarder or two, as is common in these types of open public spaces (Secchi and Viganò, 2009). It is foremost a visually striking space, and not *lived* in the same way.

Secchi and Viganò published their plan for the city in a book titled *Antwerp, territory of a new modernity* (2009). The reference to the urbanism of modernity in the title is visually recalled in Theaterplein through the negative space of Vogelenmarkt. Without the market, the space is open-ended and unbounded, as reflected in Secchi and Viganò's name for the project: *Spazio s-misurato* (Immeasurable space). Parham says that the urbanism of modernity rejects place as a vestige of the pre-modern and that 'urban redevelopment schemes on modernist principles identified markets with an outmoded past' (2015, p. 81). Two of the most renowned practitioners of this type of urbanism are Baron Haussmann, who radically reshaped Paris into the 'capital of modernity' in the mid-nineteenth century (Harvey, 2005), and the Swiss architect Le Corbusier. Haussmann and Le Corbusier may not have shared aesthetic or political objectives per se, but Le Corbusier's modernist visions for cities such as Rio de Janeiro (1929), Algiers (1933) and Addis Ababa (1936) are not unsympathetic to Haussmann's approach. Le Corbusier, like Haussmann, approached urbanism as a monosemous narrative for centralizing and mastering heterogeneous urban flows (Ten Hoor, 2007). In Le Corbusier's unrealized *Plan voisin* for Paris (1925), the city is a machine relentlessly working to sort and manage these flows (Langer, 1984). In his vertically laminate city, pedestrians have an unstriated network of pathways on a level above the similarly unimpeded circulating traffic. Modernity supplied the tools to construct this urban landscape, which Le Corbusier declared, 'exalts us with the joy that architecture provokes..., the pride which results from order, [and] the spirit of initiative which is engendered by wide spaces' (1925, n.p.). He described the *Plan voisin* as 'an actual town-planning and architectural project which has been based on concrete statistics, the proved reliability of certain materials, a new form of social and economic organization, and a more rational exploitation of real property' (1925, n.p.).

The disciplined and calculable urbanism of the modernist city was the spatial and social antidote to place, which for Le Corbusier was epitomized by the disorderly and unsanitary street:

> The street is no more than a trench, a deep cleft, a narrow passage... The street consists of a thousand different buildings, but we have got used to the beauty of ugliness for that has meant making the best of our misfortune. Those thousand houses are dingy and utterly discordant one with another. It is appalling, but we pass on our way. On Sundays, when they are empty, the streets reveal their full horror. But except during those dismal hours men and women are elbowing their way along them, the shops are ablaze, and every aspect of human life pullulates throughout their length. Those who have eyes in their heads can find plenty to amuse them in this sea of lusts and faces. It is better than the theatre, better than what we read in novels.
>
> (Le Corbusier, 1925, n.p.)

The suspicion of the street in Haussmann's designs was political and strategic: the street was the site of popular insurrection. Le Corbusier cannot decide what he finds most distasteful about the place of the street. Is its lack of an overarching plan? The inefficient and spontaneous uses to which it is put? Or, the undisciplined people who inhabit it? Is it its hybridity of form and function? Le Corbusier's queasiness about the improvisations of urban place, what Jane Jacobs (1961) celebrated as 'the ballet of the sidewalk', enjoined him to atomize the many functions of the city so they could not contaminate each other.

In Haussmann's reshaped Paris, Les Halles was a node for the flows of capitalist exchange first, and a place, second. Victor Baltard's iron and glass temples to capitalism and modernity were an encapsulation and a reflection of this. For Le Corbusier, space took precedence over place in the urban order, and given his recoil from the place of the street, it is hard to imagine that he was keen to encourage, through his urbanism, making that produced places like markets. Jacobs, in contrast, believed that the making of place from urban space was the motivation for and the pleasure of city living. In her book *The death and life of great American cities* Jacobs makes little mention of markets as urban places, but she was enthusiastic about one project that proposed using street vendors in a plaza to resurrect a moribund retail precinct. She stated that 'Deliberate street arrangements for vendors can be full of life, attraction and interest, and because of bargains are excellent stimulators of cross-use. Moreover, they can be delightful-looking' (Jacobs, 1961, p. 312).

Urbanism as placemaking

In *Antwerp, territory of a new modernity*, Secchi and Viganò occupy a terrain somewhere between Le Corbusier and Jacobs. They allude to an urbanism that creates a grand narrative for the city, while at the same time signalling a development of that approach through the 'new' in their title. Secchi and Viganò state that the emergence of place from their design for Theaterplein will be a key determinant for assessing the success of their form of urbanism.

> On one hand *Theaterplein*'s role as support for cultural life was important. On the other it needed to be transformed into a significant cityscape: a high-quality, sustainable, functional, and future-oriented space – a place that would invite people to remain there, that would interact with its surroundings… without ignoring existing uses.
>
> (Secchi and Viganò, 2009, p. 191)

The urbanism of Theaterplein activates urban space through design that neither inhibits extant processes of making place, nor fetishizes place as many new urbanisms do. The 'new' in Secchi and Viganò's urbanism, anchored as it is by 'modernity', is therefore not an affirmation of urbanisms closely aligned with placemaking, such as tactical urbanism.

Mike Lydon and Anthony Garcia, the principals of placemaking consultancy the Street Plans Collaborative, describe tactical urbanism as

> an approach to neighbourhood building and activation using short-term,
> low-cost, and scalable interventions and policies. Tactical Urbanism is
> used by a range of actors, including governments, business and nonprof-
> its, citizen groups, and individuals. It makes use of open and iterative
> development processes, the efficient use of resources, and the creative
> potential unleashed by social interaction.
>
> (Lydon and Garcia, 2015, p. 2)

Lydon and Garcia have written a manual for tactical urbanism, a format that
congeals tactics into standardized practice. This standardization is antithetical to de Certeau's notion of tactics (1984), from whom Lydon and Garcia
are aware they are appropriating. Lydon and Garcia gloss the disjuncture by
arguing that their 'empowered' tactics are distinct from de Certeau's, who
they say characterized tactics as the tool of the powerless urban citizen in a
futile struggle for the right to the city (Lydon and Garcia, 2015, pp. 9–10). In
tactical urbanism, the right to the city is repurposed as a creative right to the
city. However, Lydon and Garcia are clear that tactical urbanism is different
to other 'creative' urbanisms such as everyday, guerrilla or DIY urbanism,
which are 'the expression of the individual, or at most a small group of actors'
(2015, p. 8). Tactical urbanism can be deployed

> by municipal departments, government, developers, and nonprofit organ-
> izations to test ideas or enact change without delay. Although these ini-
> tiatives often begin with smaller citizen advocacy efforts, the benefits of
> Tactical Urbanism become clearer as they are integrated into the munici-
> pal project delivery process and capably brought to neighborhoods across
> the city.
>
> (Lydon and Garcia, 2015, p. 8)

In doing so, tactical urbanism is 'usually intended to instigate long-term
change, such as revising an outdated policy or responding to a deficiency of
infrastructure' (Lydon and Garcia, 2015, p. 8).

Lydon and Garcia propose long-term change in cities be achieved through
temporary initiatives that they call 'placeholder projects' (2015, p. 16). The
rationale for this is that 'the places people inhabit are never static, Tactical
Urbanism doesn't propose one-size-fits-all solutions but intentional and flex-
ible responses' (2015, p. 3). Here, place is malleable and modular, and therefore able to be configured according to convenience, expediency or fashion. In
placemaking, place can be constructed and customized to order.

Given this insistence on temporariness it is no coincidence that the 'pop-up'
is a key element in tactical urbanism (Mould, 2015). The value of markets as
placemaking devices for Lydon and Garcia is their pop-up potential: 'Street

fairs and bazaars, markets, block parties, and similar temporary events have brought life to streets for millennia, proving that our thoroughfares fulfil a rich social and economic purpose as much as a utilitarian one' (2015, p. 39). In this way, markets are grouped with 'food trucks, pop-up stores, better block initiatives, chair bombing, parklets, shipping container markets, do-it-yourself (DIY) bike lanes, guerrilla gardens, and other hallmarks of the Tactical Urbanism movement' (Lydon and Garcia, 2015, p .6). Markets appeal as a device for placemaking precisely because of their association with place. Lydon and Garcia's consulting firm commissioned a report (Flynn, n.d.) on how market culture in South America could be transposed for the 'revival' of markets in the US as sites of placemaking-led entrepreneurial activity. Another global leader in the field, Project for Public Spaces, has a dedicated markets 'division', and organises workshops on how to run successful markets (Public Markets, n.d.).

Placemaking positions itself as the creative, alternative or subversive counterpoint to car-centric uses of urban space by advocating for everyday practices such as walking or spending time in public space. At times, Lydon and Garcia's manual reads like a mash-up of de Certeau and Unitary Urbanism, the Situationists' radical rejection of the urbanism of modernity (Kotányi and Vaneigem, 1961). Situationism advocated a decentred approach to urbanism, which deployed chance, digression, improvisation, spontaneity and the ludic in aleatory cartographies and negotiations of urban space such as the *dérive* and the *détournement*. Above all, the *dérive* and the *détournement* should resist and/or subvert dominant existing spatial formation and organization, as this anonymous tract from the *Situationiste Internationale* explains:

> UU [Unitary Urbanism] is not ideally separated from the current terrain of cities. UU is developed out of the experience of this terrain and based on existing constructions. As a result, it is just as important that we exploit the existing decors through the affirmation of a playful urban space such as is revealed by the *dérive* as it is that we construct completely unknown ones. This interpenetration (employment of the present city and construction of the future city) entails the deployment of architectural *détournement*.
>
> (Anonymous, 1959, n.p.)

Contemporary urbanisms, in theory and in practice, tend towards Jacobs' recognition of the pleasures and benefits of urban place, and do so by acknowledging the heterogeneous and global flows of the city within a localized, place-based framework. Yet in the face of totalizing spatializations such as the hegemony of automobility, surveillance technology and neo-liberal commodification of public space, the artefacts of placemaking, like deckchairs or food trucks, tend to distract from serious objectives like environmental sustainability and the realization of the human scale in alienating urban landscapes. Urbanist impulses are channelled into creating small, contained areas for certain type of consumers – ones that value ethical consumption, sustain-ability and 'creativity'– as opposed to producing sustainable and everyday urbanism that is impactful for a wider cross-section of the city than the social groups who are already well-served by these types of urban interventions. Regardless of Lydon and Garcia's declarations to the contrary, placemaking and its ally Creative City discourse often propagate one-size-fits-all solutions that are not always the most appropriate or equitable in managing the chal-lenges of urbanization. When Mexico City's government announced a plan to build its own version of New York's renowned and very successful High Line park (Bliss, 2015), critics pointed out that the project proposed a privately operated park and mall as a solution to improved access to public transport, when the far less expensive option of some traffic lights and pedestrian cross-ings would be equally effective. Furthermore, the project is located in the cen-tre of Mexico City, which already attracts the majority of public and private investment in urban development. This placemaking project comes out of the type of urban policy and planning that Mould says is

> designed less to tackle the root causes of many difficult social, cultural and economic challenges that [cities] face via cultural and creative means, and more to excuse and justify activities that promote and valorise eco-nomic production and profit-making for interested private (and public) stakeholders at the expense of everything else.
>
> (Mould, 2015, p. 17)

The human scale of markets

Abdallah Khatib's (1958) Situationist psycho-geography, to which I have referred in previous chapters, noted how Les Halles created an urbanism by virtue of it being a place that was less hostile to pedestrians than the rest of Paris' streets. Today, Khatib's project would surely be appropriated by tactical urbanism as one of its own. In Lydon and Garcia's genealogy of tactical urbanism, the movement is a response to the hegemony of the car because 'almost as soon as cars began dominating urban streets, tactical interventions were organized to take them back, even if only temporarily' (2015, p. 39). Khatib wrote, 'despite appearances the quarter of Les Halles is one of the easiest to cross, via the access routes which border or cross it in every direction' (1958, n.p.). His discovery at Les Halles is what urbanists call the 'human scale'. Jan Gehl, who was instrumental in the people- and place-centred re-design of central Copenhagen (Gehl & Gemzøe, 2004), is a key proponent of the human scale in urban design. Designing to and for the human scale is one response to a situation in contemporary cities where walk-ing or congregating in public is marginalized to the extent that these practices are considered resistant or subversive. Gehl's book *Cities for people* (2010) is a comprehensive study of the social, spatial, sensory and material dimensions of the human scale. Rather than surveying the city purely with a panoramic (and panoptic [Foucault, 1977]) eye – as de Certeau (1984) did from the top of the World Trade Center – the human scale shifts perspective and attention to what is at eye-level, and to the micro-climates of place in the city.

Like the Situationists, Gehl is a critic of the urbanism of modernism, which, along with the hegemony of automobility, he holds jointly responsible for the 'shattered scale' of contemporary cities (2010, p. 55). Together, he insists, they thwart the possibilities for sociality and encounter that are vital for urban life. For Gehl, the human scale is a way of (re)folding the social into the fabric of cities (Gehl, 2010; Sørensen and Dalsgaard, 2012). One way that urbanists have rejected the panoramic scale of modernist urbanism and signalled a return to the human scale of place is by evoking the image of the village within the city. In The structure plan for Antwerp, Secchi and Viganò (citing Rasmussen, 1982 [1934]) conceptualize Antwerp as both 'a city of villages and a metropolis' (2009, p. 123). The 'city of villages' has proved an enduring model for thinking about urban place (Oldenburg, 1999; City of Sydney, 2011b; Little book of Sydney villages, n.d.). The recourse to the village is understandable as a conceptual and spatial means of grappling with the incomprehensible scale of urbanization (Tonkiss, 2013; Merrifield, 2014). Creating marketplaces is specifically identified by Gehl (2010) as a way of building human scale into urban space. In Young et al.'s 'criteria for evaluating potential farmers' markets sites' in Philadelphia (2011, p. 209), many of the favourable elements on their checklist are representative of the human scale as outlined by Gehl, including visibility of location; potential as a gathering space; pedestrian traffic; walkability; bike access; public transportation; attractive space (shade, trees, benches, sidewalks); safety (lighting, sidewalks, traffic-calming measures, sightlines); nearby retail locations (other non-food retail outlets); and other amenities (recreation opportunities, bike racks, water fountains, restrooms, trash cans).

Markets create 'soft edges' (Gehl, 2010, p. 75), which are interstitial spaces that transition between built and social environments. Vogelenmarkt's 'softening' of Theaterplein is an example of this. Markets achieve this because they are 'congregant spaces' that 'cater to the human desire for sociability, for experiencing the presence of other humans, bumping into friends and acquaintances, enjoying the hum and babble of the crowd' (Davison, 2006, p. 1). Pierre Mayol, in his research into everyday life, stressed the importance of the social space of the market as a means of understanding urban experience: 'The market is traditionally an important sociological landmark for the understanding of human relations within the practice of the neighbourhood. No city or village is without one' (de Certeau et al., 1998, p. 107). Stuart Plattner (1982) observed that social networks and competencies are as crucial to efficient operations and management in markets as any other skill set or technological and material infrastructure. Sociality is by no means exclusive to markets amongst urban spaces and places. In writing about London's Borough Market, Coles and Crang (2011) situate markets within a spectrum of retail and consumer spaces where sociality is a factor. Given that Borough

Market is a high-end foodie market that does a roaring trade in expensive, take-away lunches for workers in nearby office blocks like The Shard, it is unsurprising that the sociality there was no different to that to be found in other consumer spaces. In everyday markets, however, Watson and Studdert (2006) found a sociality that was distinctive in its inclusiveness and hetero-geneity. In their research on local markets in the UK, they noted that they are 'a public space where marginalised groups come to spend time, thereby pro-viding opportunities to escape isolation in the home or elsewhere, while also providing an economically inclusive space – for example, by offering cheap goods that may not be available elsewhere' (2006, p. 14). In globalized cities, it is hard to find many other equivalent places in which to freely loiter, chat and hang about. Markets can provide social opportunities, but also the possibility of participating in consumer culture for those with limited possibilities to do so otherwise.

Learning from markets

In 2014, not far from Antwerp, the city of Rotterdam opened Markthal, a complex of apartments and a vast parking space wrapped around a market-place. Its architects, MVRDV, describe the project as 'a sustainable combi-nation of food, leisure, living and parking, fully integrated to celebrate and enhance the synergetic possibilities of the different functions' (Market Hall, n.d.). It is statement architecture and barely a year on, it has won a fistful of architectural and design awards. Like Secchi and Viganò's design for Theaterplein, Rotterdam's new market, built at a cost of more than €175 mil-lion (Bevan, 2015), sends the message that markets are worthy of serious architecture, and signals their value to the urban economy.

In many ways, Markthal is an allegory about the future of markets as they are being discursively situated in globalized cities. It is an example of a mar-ket being deployed as vehicle for urban renewal, like Carriageworks Farmers Market in Sydney, or Enric Miralles and Benedetta Tagliabue's 2005 remod-elling of Mercat Santa Caterina in Barcelona, which was a design influence for Markthal (Bevan, 2015). One of the principal architects, Winy Maas, said in an interview (Wainwright, 2014) that he hoped that the market would be the impetus for future market concretion and colonization in Rotterdam's city centre (Conzen and Conzen, 2004). Markethal is a privately developed facility, whereby the sale or monetization of public land offsets the cost of the renewal project and supposedly ensures the 'viability' of the public assets left over. Its design is inscribed with the Creative City – literally – with a site-specific digital artwork (the largest of its kind in the Netherlands) lining the interior. Essentially, Markethal is a themed environment. Shoppers enjoy the spec-tacle of 'a foodie utopia of artisan bakers and biological butchers, aimed at a more upmarket clientele', and then head to Blaak Markt, the cheaper mar-ket just outside in what is left of the square, which 'accommodates a broad church, from fine fishmongers to bric-a-brac junk' (Wainwright, 2014, n.p.).

Wainwright comments that 'despite [Markethal] being a beacon of downtown urban life, it has a strangely suburban approach to planning', as evidenced by the 'gigantic chain supermarket in the basement below' and the construction 'deep within its bowels... [of] the city's biggest car park, with 1,200 spaces, despite the metro and bus station being right opposite' (2014, n.p.). Bevan's (2015) assessment was that the final effect was more food court than market.

Markthal is emblematic of the global renaissance that markets are undergoing after a long steady decline in the wake of mass consumer culture (Visconti et al., 2014). These emergent and reconfigured markets are being shaped by and contributing to globalized discourses of place, and are therefore much more self-conscious iterations of traditional markets (Parham, 2015). The emergence of highly visible, spectacular markets like Markethal may seem like a positive step for market culture in cities, but it is problematic. Growing numbers of privatized, specialist markets are endemic of the commodification of public spaces that undermines markets' enduring status as civic places. Markets are affected by concomitant neo-liberal disciplining and marginalization of public space and place. Any emphasis on improvement and amenity of public space in gentrified or regenerated areas is, ironically, motivated by the undervaluing of public places and public practices which views them solely as a means of capturing surplus value. Examples from Rio de Janeiro, Hong Kong and Beijing in this book demonstrate that when cities are 'modernized', 'streamlined', 'regenerated' or 'cleaned up' to become globalized cities, street markets and vendors are among the first moved off the streets.

In cities where markets are part of public infrastructure, for example, as health initiatives and community development (as in San Francisco), or as vernacular consumer culture (as in Antwerp and Paris), inclusive markets are surviving. Everyday markets also continue to survive where informality is a socially accepted part of the market landscape. Cities that support everyday markets through policy, planning, and urban design are supporting urban sociality (Watson, 2009), the revalorization of ordinary places (Knox, 2005) and an urban fabric that emerges from heterogeneous forms and ways of making. Returning to the city as bazaar, the metaphor that I introduced at the beginning of the book, it strikes me that there are worse ways to imagine the city. The city as bazaar presents possibilities for an urbanism that produces subaltern, informal, ordinary, collective places in the city. Markets produce an urbanism of the human scale that generates place through making such as quotidian provisioning, hanging about, chatting and walking. This is making that is produced from and valorizes the rhythms of everyday urban life. The making that happens in these markets is creative and improvised and makes possible the right to the city that Lefebvre envisaged.

Urbanists, urban designers, planners, architects and policy-makers can learn a lot on how to build cities that are place-based from these types of markets. Gehl has already recognized that markets are one way to achieve the human scale (2010; see also Sørensen and Dalsgaard, 2012). CittàSlow, the international organization advocating slow urbanism, has also recognized that markets 'propagate vitality ... and the creation of inviting public spaces' (Knox, 2005, p. 7) in towns and cities. Tigran Haas, a proponent of sustainable urbanism, says that cities 'will be (and are) both the problem and the solution for solving urban problems' (2012, p. 9). This statement may read as oxymoronic, but it raises the possibility that the solutions to the challenges of urbanization are within the urban fabric itself. Markets, a key urban form, can be one of these solutions. It is those markets that are truly urban, in the sense that Lefebvre understood the urban – everyday, heterogeneous, liminal, improvisational – that are best positioned to offer grounded and creative responses. Unfortunately, it is these types of markets that are under threat in globalized cities. Therefore when we ask ourselves, 'What kind of cities do we want?' or 'What kind of places do we want to make in our cities?', we should also be asking, 'What kind of urban markets do we want?'

Bibliography

4e site touristique de France (n.d.). Retrieved from http://marcheauxpuces-saintouen. com/3.aspx.

ABC (2004). Redeveloping The Block. *7.30 Report*, 12 December. Retrieved from www.abc.net.au/7.30/content/2004/s1267209.htm.

Abdulazeez, Y. and Pathmanathan, S. (2014). Migrants in informal urban street marekts: experience from Sokoto. In C. Evers and K. Seale (Eds.), *Informal urban street markets: international perspectives* (pp. 37–50). New York: Routledge.

About our market. (n.d.) Retrieved from http://heartofthecity-farmersmar. squarespace.com/about.

Agamben, G. (1993). *The coming community*. Minneapolis: University of Minnesota Press.

Agnew, J.-C. (1986). *Worlds apart: the market and the theatre in Anglo-American thought, 1550–1750*. Cambridge: Cambridge University Press.

AHC (n.d.). Pemulwuy. Retrieved from www.ahc.org.au/pemulwuy.

Alameda Point Antiques Faire (n.d.). Retrieved from www.yelp.com.au/biz/alameda-point-antiques-faire-alameda.

Alkon, A. H. (2012). *Black, white, and green: farmers markets, race, and the green economy*. Athens: University of Georgia Press.

Alkon, A. H. and Agyeman, J. (2011). *Cultivating food justice: race, class, and sustainability*. Cambridge, MA: MIT Press.

Alkon, A. H. and McCullen, C. G. (2011). Whiteness and farmers markets: performances, perpetuations… contestations? *Antipode*, 43(4), 937–959.

Anderson, B. (2009). Affective atmospheres. *Emotion, Space and Society*, 2(2), 77–81.

Anderson, E. (2011). *The cosmopolitan canopy: race and civility in everyday life*. New York: W.W. Norton & Co.

Anderson, K. (1993). Place narratives and the origins of Inner Sydney's Aboriginal settlement, 1972–73. *Journal of Historical Geography*, 19(3), 314–335.

Anholt, S. (2010). *Places: identity, image and reputation*. Basingstoke: Palgrave Macmillan.

Anonymous (1959). Unitary Urbanism at the end of the 1950s. *Internationale Situationniste, 3*. Retrieved from www.cddc.vt.edu/sionline/si/unitary.html.

Arbaci, S. and Tapada-Berteli, T. (2012). Social inequality and urban regeneration in Barcelona city centre: reconsidering success. *European Urban and Regional Studies*, 19(3), 287–311.

Aspers, P. (2011). *Markets*. Cambridge: Polity.

Augé, M. (1995). *Non-places: introduction to an anthropology of supermodernity.* London: Verso.

Augoyard, J.-F. (1998). Eléments pour une théorie des ambiances architecturales et urbaines. *Les Cahiers de la recherche architecturale,* 42/43(3), 7–23.

Augoyard, J.-F. (2007). *Step by step: everyday walks in a French urban housing project.* Minneapolis: University of Minnesota Press.

Bailey, M. (2010). Inside suburban 'Persian bazaars': the reception of regional shopping centres in Sydney during the 1960s. In R. Crawford, J. Stuart and K. Humphery (Eds.), *Consumer Australia: historical perspectives* (pp. 119–134). Newcastle: Cambridge Scholars.

Bakhtin, M. (1984). *Rabelais and his world.* Bloomington, IN: Indiana University Press.

Bakhtin, M. (1986). Forms of time and of the chronotope in the novel. In C. Emerson and M. Holquist (Eds.), *Speech genres and other late essays* (pp.84–258). Trans. Vern W. McGee. Austin: University of Texas Press.

Baldwin, S. (2011). Stall tales. *Jamie Magazine,* 23, 51–62.

Barber, L. (2014). (Re)making heritage policy in Hong Kong: a relational politics of global knowledge and local innovation. *Urban Studies,* 51(6), 1179–1195.

Barry, A. and Slater, D. (2002). Technology, politics and the market: an interview with Michel Callon. *Economy and Society,* 31(2), 285–306.

Bauman, Z. (2004). *Wasted lives: modernity and its outcasts.* Oxford: Polity.

Bauman, Z. (2011). *Collateral damage: social inequalities in a global age.* Cambridge: Polity.

Begg, Z. and Stickells, L. (Eds.) (2011). *The right to the city.* Sydney: Tin Sheds Gallery, Faculty of Architecture, Design and Planning, University of Sydney.

Behrendt, L. (2006). What lies beneath: how traditional aboriginal values, effectively harnessed to contemporary social policy-making, can help invigorate Australian urban communities. *Meanjin,* 65, 4–12.

Bela, J. (2014). User-generated urbanism and the right to the city. In J. Hou, B. Spencer, T. Way and K. Yocom (Eds.), *Now urbanism: the future city is here* (pp. 149–164). Abingdon: Routledge.

Bell, J. S. and Loukaitou-Sideris, A. (2014). Sidewalk informality: an examination of street vending regulation in China. *International Planning Studies,* 19(3–4), 221–243.

Benedetta, M. and Moholy-Nagy, L. (1972 [1936]). *The street markets of London.* New York: B. Blom.

Benjamin, W., and Tiedemann, R. (1999). *The arcades project.* Trans. H. Eiland and K. McLaughlin. Cambridge, MA, and London: Belknap Press.

Benton, M. (Producer) (1985). Incomers [Television series episode]. In *Ours to keep.* Bristol: BBC.

Berman, M. (1983). *All that is solid melts into air: the experience of modernity.* London: Verso.

Bevan, R. (2015). Vegetal vault. *Architectural Review,* 237(1415), 56–67.

Bhabha, H. K. (1994). *The location of culture.* London: Routledge.

Black, R. (2012). *Porta Palazzo: the anthropology of an Italian market.* Philadelphia: University of Pennsylvania Press.

Blake, M. (2014). Stifling street life: the demise of Graham Street Market in Hong Kong [blog post], 10 April. Retrieved from http://geofoodie.org/2014/04/10/stifling-street-life-the-demise-of-graham-street-market-in-hong-kong.

Bliss, L. (2015). The backlash to Mexico City's High Line-style park, 29 September. Retrieved from www.citylab.com/design/2015/09/the-terrible-plan-for-mexico-citys-high-line-style-park/408010.

Blommaert, J. (2013). *Ethnography, superdiversity and linguistic landscapes: chronicles of complexity*. Bristol: Multilingual Matters.

Böhme, G. (1993). Atmosphere as the fundamental concept of a new aesthetics. *Thesis Eleven*, 36, 113–126.

Böhme, G. (2013). The art of the stage set as a paradigm for an aesthetics of atmospheres. *Ambiances: International Journal of Sensory Environment, Architecture and Urban Space*. Retrieved from http://ambiances.revues.org/315.

Bondi Junction Village Markets (2015). Retrieved from www.localmarketguide.com.au/295-bondi-junction-village-markets#.Vcb2Mp2qqko.

Bondi Junction: key stats and figures (n.d.) Retrieved from www.scentregroup.com/centre/westfield-bondi-junction.

Bourdieu, P. (1984). *Distinction: a social critique of the judgement of taste*. London: Routledge and Kegan Paul.

Bové, J. and Dufour, F. (2001). *The world is not for sale: farmers against junk food*. London: New York: Verso.

Bové, J. and Dufour, F. (2005). *Food for the future: agriculture for a global age*. Cambridge: Polity Press.

Brandão, Z. (2006). Urban planning in Rio de Janeiro: a critical review of the urban design practice in the twentieth century. *City & Time*, 2(2). Retrieved from www.ceci-br.org/novo/revista/rst/viewarticle.php?id=53.

Braudel, F. (1981). *The structures of everyday life: the limits of the possible*. Trans. S. Reynolds. London: Collins.

Braudel, F. (1982). *The wheels of commerce*. Trans. S. Reynolds. London: Collins.

Breman, J. (2013). *At work in the informal economy of India: a perspective from the bottom up*. Oxford: Oxford University Press.

Broudehoux, A.-M. (2004). *The making and selling of post-Mao Beijing*. New York and London: Routledge.

Broudehoux, A.-M. (2007). Spectacular Beijing: the conspicuous construction of an Olympic metropolis. *Journal of Urban Affairs*, 29(4), 383–399.

Brown, B. (2003). *A sense of things: the object matter of American literature*. Chicago: University of Chicago Press.

Bryson, N. (1997). Mark Dion and the birds of Antwerp. In L. G. Corrin, M. Kwon and N. Bryson (Eds.), *Mark Dion* (pp. 88–97). London: Phaidon.

Bubinas, K. (2011). Farmers markets in the post-industrial city. *City & Society*, 23(2), 154–172.

Calabi, D. (2004). *The market and the city: square, street and architecture in early modern Europe*. Aldershot: Ashgate.

California Code of Regulations. Retrieved from https://govt.westlaw.com/calregs/Index?transitionType=Default&contextData=(sc.Default)

Carter, P. (1987). *The road to Botany Bay: an essay in spatial history*. London: Faber.

Cartier, C. (2013). Class, consumption and the economic restructuring of consumer space. In M. Chen and D. S. G. Goodman (Eds.), *Middle class China: identity and behaviour* (pp. 34–53). Cheltenham: Edward Elgar.

Casson, M., and Lee, J. S. (2011). The origin and development of markets: a business history perspective. *Business History Review*, 85(1), 9–37.

Castells, M. (1996). *The rise of the network society*. Oxford: Blackwell.

Certified Farmers Markets Program (n.d.) Retrieved from www.cdfa.ca.gov/is/i_&_c/cfm.html.

Chang, T. C. and Teo, P. (2009). The Shophouse Hotel: vernacular heritage in a Creative City. *Urban Studies*, 46(2), 341–367.

Chevalier, L. (1994). *The assassination of Paris*. Chicago and London: University of Chicago Press.

Christie, M. (1988). *The Sydney markets: 1788–1988*. Sydney: Sydney Markets Authority.

City of Sydney (2011a). Redfern–Waterloo Draft Built Environment Plan 2 [BEP2]: Submission to the Redfern–Waterloo Authority. Retrieved from www.cityofsydney. nsw.gov.au/development/UrbanRenewalProjects/Default.asp.

City of Sydney (2011b). *Slices of Sydney: a taste of our villages*. Sydney: Urban Walkabout.

Clark, G. (2010). *Onions are my husband: survival and accumulation by West African market women*. Chicago: University of Chicago Press.

Coleman, S. and Crang, M. (Eds.) (2002). *Tourism: between place and performance*. New York and Oxford: Berghahn.

Coles, B. and Crang, P. (2011). Placing alternative consumption: commodity fetishism in Borough Fine Foods Market, London. In T. Lewis and E. Potter (Eds.), *Ethical consumption: a critical introduction* (pp. 87–102). London: Routledge.

Coletto, D. (2010). *The informal economy and employment in Brazil: Latin America, modernization, and social changes*. New York: Palgrave Macmillan.

Community (n.d.). Retrieved from http://broadwaymarket.co.uk/community.php.

Conzen, M. R. G. and Conzen, M. P. (2004). *Thinking about urban form: papers on urban morphology, 1932–1998*. Bern and Oxford: Peter Lang.

Conzen, M. and Larkham, P. (Eds.) (2014). *Shapers of urban form: explorations in morphological agency*. New York: Routledge.

Cooke, R. (2009). Iain Sinclair: the interview. *Guardian*, February 8. Retrieved from www.guardian.co.uk/books/2009/feb/08/iain-sinclair-interview.

Crawford, R., Stuart, J. and Humphery, K. (2010). *Consumer Australia: historical perspectives*. Newcastle: Cambridge Scholars.

Cross, J. C. and Morales, A. (2007). *Commerce in a globalizing world: street sales*. London: Routledge.

Cummings, J. (2015). Confronting favela chic: the gentrification of informal settlements in Rio de Janeiro, Brazil. In L. Lees, H. B. Shin and E. López Morales (Eds.), *Global gentrifications: uneven development and displacement* (pp. 81–100). Bristol: Polity Press.

da Cunha, N. V. and de Mello, P. P. T. (2014). Rio de Janeiro's global bazaar: Syrian, Lebanese, and Chinese merchants in the Saara. In P. Amar (Ed.), *The Middle East and Brazil: perspectives on the new global south* (pp. 228–240). Bloomington, IN: Indiana University Press.

da Matta, R. (1991). *Carnivals, rogues, and heroes: an interpretation of the Brazilian dilemma*. Trans. J. Drury. Notre Dame, IN: University of Notre Dame Press.

Davison, G. (2006). From the market to the mall: a short history of shopping in Melbourne. Background Report, Victorian Retail Policy Review, Department of Planning and Community Development.

De Bruin, A. and Dupuis, A. (2000). The dynamics of New Zealand's largest street market: the Otara Flea Market. *The International Journal of Sociology and Social Policy*, 20(1–2), 52–73.

de Certeau, M. (1984). *The practice of everyday life*. Berkeley and London: University of California Press.

de Certeau, M., Giard, L. and Mayol, P. (1998). *The practice of everyday life. Vol.2, living and cooking*. Minneapolis: University of Minnesota Press.

Deleuze, G. and Guattari, F. L. (1987). *A thousand plateaus: capitalism and schizophrenia*. Minneapolis: University of Minnesota Press.

del Rio, V. and de Alcantara, D. (2009). The cultural corridor project: revitalization and preservation in downtown Rio de Janeiro. In V. del Rio and W. J. Siembieda (Eds.), *Contemporary urbanism in Brazil: beyond Brasilia* (pp. 125–143). Gainesville: University Press of Florida.

de Kretser, M. (2009). Odds and endings. *The Age*, August 29, p. 16.

Dines, N. (2009). The disputed place of ethnic diversity: an ethnography of the redevelopment of a street market in East London. In R. Imrie, L. Lees and M. Raco (Eds.), *Regenerating London: governance, sustainability and community in a global city* (pp. 254–272). London: Routledge.

Dines, N. and Cattell, V. (2006). *Public spaces, social relations and well-being in East London*. Bristol: Policy Press.

Douglas, M. (1966). *Purity and danger: an analysis of concepts of pollution and taboo*. London: Routledge and Kegan Paul.

Duru, A. (2014). The politics of space in the marketplace: re-placing periodic markets in Istanbul. In C. Evers and K. Seale (Eds.), *Informal Urban Street Markets* (pp. 149–157). Abingdon: Routledge.

Dutton, M. (2008). *Beijing time*. Cambridge, MA: Harvard University Press.

Elias, N. (1978). *The civilising process: the history of manners: sociogenetic and psychogenetic investigations*. Trans. Edmund Jephcott. Oxford: Basil Blackwell.

Entrikin, J. N. (1990). *The betweenness of place: toward a geography of modernity*. Baltimore, MD: Johns Hopkins University Press.

Eveleigh Farmers' Market (n.d.). Retrieved from http://web.archive.org/web/2015 0528220638/http://www.eveleighmarket.com.au.

Evers, C. (2014). Pengpu Night Market: informal urban street markets as more-than-human assemblages in Shanghai. In C. Evers and K. Seale (Eds.), *Informal urban street markets: international perspectives* (pp. 95–104). New York: Routledge.

Evers, C. and Seale, K. (2014) *Informal urban street markets: international perspectives*. New York: Routledge.

Farr, D. (2008). *Sustainable urbanism: urban design with nature*. Hoboken, NJ: John Wiley.

Feld, S. (1996). Waterfalls of song: an acoustemology of place resounding in Bosawi, Papua New Guinea. In S. Feld and K. Basso (Eds.), *Senses of place* (pp. 91–136). Santa Fe: School of American Research Press.

Félix, F. (2002). Saint-Ouen: une ZPPAUP qui maintient les Puces. *Le moniteur*, January 4. Retrieved from www.lemoniteur.fr/articles/saint-ouen-une-zppaup-qui-maintient-les-puces-304459.

Fitzgerald, S. (1997). *Red tape, gold scissors: the story of Sydney's Chinese*. Sydney: State Library of New South Wales Press.

Fligstein, N. and Dauter, L. (2007). The sociology of markets. *Annual Review of Sociology*, 33, 105–128.

Florida, R. (2002) *The rise of the creative class*. New York: Basic Books.

Flynn, J. (n.d.). *Mercado: lessons from 20 markets across South America*. Miami and New York: The Streets Plan Collective.

Foley, G. (2001). Black power in Redern 1968–1972. The Koori history website. Retrieved from www.kooriweb.org/foley/essays/essay_1.html.

Foley, G. (2012). *Foley*. Live performance at the Playhouse, 24 January, Sydney Opera House, Sydney.

Foley, G., Schaap, A. and Howell, E. (Eds.) (2014). *The Aboriginal tent embassy: sovereignty, black power, land rights and the State.* Abingdon: Routledge.

Food deserts (n.d.). Retrieved from apps.ams.usda.gov/fooddeserts/fooddeserts.aspx.

Foucault, M. (1973). *The order of things: an archaeology of the human sciences.* London: Vintage.

Foucault, M. (1977). *Discipline and punish: the birth of the prison.* Trans. Alan Sheridan. London: Penguin.

Foucault, M. and Miskowiec, J. (1986). Of other spaces. *Diacritics*, 16(1), 22–27.

Franck, K. A. (2005). The city as dining room, market and farm. *Architectural Design*, 75(3), 5–10.

Freire-Medeiros, B. (2013). *Touring poverty*. London: Routledge.

Freitas, M. (2013). Fim da Perimetral não extingue tradição da Feira da Praça XV. Retrieved from http://puc-riodigital.com.puc-rio.br/Jornal/Cidade/Fim-da-Perimetral-nao-extingue-tradicao-da-Feira-da-Praca-XV-23698.html#.U_QX1PbXYl8.

Frenzel, F., Steinbrink, M. and Koens, K. (2012). *Slum tourism: poverty, power and ethics.* New York: Routledge.

Fuller, G., Hamilton, C. and Seale, K. (2013). Working with amateur labour: between culture and economy. *Cultural Studies Review*, 19(1), 143–154.

Gammage, B. (2011). *The biggest estate on earth: how Aborigines made Australia.* Crows Nest: Allen & Unwin.

Geertz, C. (1963). *Peddlers and princes; social change and economic modernization in two Indonesian towns.* Chicago: University of Chicago Press.

Geertz, C. (1978). The bazaar economy: information and search in peasant marketing. *The American Economic Review*, 68(2), 28–32.

Gehl, J. (2010). *Cities for people.* Washington, DC: Island Press.

Gehl, J. and Gemzøe, L. (2004). *Public spaces, public life, Copenhagen* (3rd edn). Copenhagen: Danish Architectural Press and the Royal Danish Academy of Fine Arts, School of Architecture Publishers.

Gibson-Graham, J. K. (2006). *A postcapitalist politics.* Minneapolis: University of Minnesota Press.

Gilroy, P. (2004). *After empire: melancholia or convivial culture?* London: Routledge.

Glass, R. (Ed.) (1964). *London: aspects of change.* London: Macgibbon & Kee.

Godfrey, B. J. and Arguinzoni, O. M. (2012). Regulating public space on the beachfronts of Rio de Janeiro. *Geographical Review*, 102(1), 17–34.

Gonzalez, S. and Waley, P. (2013). Traditional retail markets: the new gentrification frontier? *Antipode*, 45(4), 965–983.

Gottlieb, R. and Joshi, A. (2010). *Food justice.* Cambridge, MA: MIT Press.

Gregson, N. and Crewe, L. (2003). *Second-hand cultures.* London: Bloomsbury.

Griffero, T. (2014). *Atmospheres: aesthetics of emotional spaces.* Farnham: Ashgate.

Guthman, J. (2011). 'If they only knew': the unbearable whiteness of alternative food. In A. H. Alkon and J. Agyeman (Eds.), *Cultivating food justice: race, class, and sustainability* (pp. 263–282). Cambridge, MA: MIT Press.

Haas, T. (2012). *Sustainable urbanism and beyond: rethinking cities for the future.* New York and London: Rizzoli.

Hackney Council (2009). Markets and street trading: an introduction to Hackney's markets. Retrieved from www.hackney.gov.uk/Assets/Documents/6-an-introduction-to-hackneys-markets.pdf.

Hackney Council (2010) Creative Hackney. Retrieved from www.hackney.gov.uk/Assets/Documents/cultural-framework.pdf.

Hackney Hipster Hate (2010). Broadway market stragglers [blog post], August 11. Retrieved from http://hackneyhipsterhate.tumblr.com/post/933106637/extra.

Hage, G. (1998). *White nation: fantasies of white supremacy in a multicultural society.* Annandale/West Wickham: Pluto Press/Comerford and Miller.

Hall, C. M. and Rath, J. (2007). Tourism, migration and place advantage in the global cultural economy. In J. Rath (Ed.), *Tourism, ethnic diversity and the city* (pp. 1–24). New York and London: Routledge.

Halls Amsterdam: Centre for Media, Fashion Culture and Crafts (2015). Retrieved from www.europanostra.org/awards/159.

Hart, K. (1973). Informal income opportunities and urban employment in Ghana. *Journal of Modern African Studies*, 11, 61–89.

Harvey, D. (2005). The political economy of public space. In S. Low & N. Smith (Eds.), *The politics of public space* (pp. 17–33). New York: Routledge.

Harvey, D. (2008). The right to the city. *New Left Review*, 53, 23–40.

Hazan, E. (2010). *The invention of Paris: a history in footsteps.* London: Verso.

Hepworth, K. (2014). Geographies of unauthorized street trade and the 'fight against counterfeiting' in Milan. In C. Evers and K. Seale (Eds.), *Informal urban street markets: international perspectives* (pp. 167–175). New York: Routledge.

Hiebert, D., Rath, J. and Vertovec, S. (2014). Urban markets and diversity: towards a research agenda. *Ethnic and Racial Studies*, 1–19.

Highmore, B. (2011). *Ordinary lives: studies in the everyday.* New York: Routledge.

Hippie Fair Crafts Market, Rio de Janeiro (n.d.) Retrieved from: http://web.archive.org/web/20130512055649/http://www.tripadvisor.com.au/Attraction_Review-g303506-d317891-Reviews-Hippie_Fair_Crafts_Market-Rio_de_Janeiro_State_of_Rio_de_Janeiro.html.

History of the flea market (n.d.) Retrieved from www.st-ouen-tourisme.com/en/decouvertes/fiche/303-histoire-des-puces.html.

Holgersson, H. (2014). Post-political narratives and emotions: dealing with discursive displacement in everyday life. In H. Jones and E. Jackson (Eds.), *Stories of cosmopolitan belonging: emotion and location* (pp. 115–126). Abingdon: Routledge.

hooks, b. (1992). *Black looks: race and representation.* Boston, MA: South End Press.

Ingold, T. (2011). *Being alive: essays on movement, knowledge and description.* Hoboken: Taylor & Francis.

Ingold, T. (2013). *Making: anthropology, archaeology, art and architecture.* Abingdon: Routledge.

Iveson, K. (2013). Cities within the city: do-it-yourself urbanism and the right to the city. *International Journal of Urban and Regional Research*, 37(3), 941–956.

Jacobs, J. (1961). *The death and life of great American cities.* New York: Random House.

Jacobs, J. M. (1996). *Edge of empire: postcolonialism and the city.* London: Routledge.

Jaguaribe, B. (2014). *Rio de Janeiro: urban life through the eyes of the city.* New York: Routledge.

Jaguaribe, B. and Salmon, S. (2012). Reality tours: experiencing the 'real thing' in Rio de Janeiro's favelas. In T. Edensor & M. Jayne (Eds.), *Urban theory beyond the West: a world of cities* (pp. 239–260). London & New York: Routledge.

James, E. (2009). *Battle for Broadway Market* [video file]. Retrieved from https:// vimeo.com/12992826.

Jansen-Verbeke, M. (1991). Leisure shopping: a magic concept for the tourism industry? *Tourism Management*, 12(1), 9–14.

Janssens, F. (2014). Street food markets in Amsterdam: unravelling the original sin of the market trader. In R. de Cássia Vieira Cardoso, M. Companion and S. R. Marras (Ed.), *Street food: culture, economy, health and governance* (pp. 98–116). Abingdon: Routledge.

Janssens, F. and Sezer, C. (2013). Marketplaces as an urban development strategy. *Built Environment*, 39(2), pp. 169–171.

Jivén, G. and Larkham, P. J. (2003). Sense of place, authenticity and character: a commentary. *Journal of Urban Design*, 8(1), 67–81.

Johnson, D. (director) (2014). *The Redfern story* [Film]. Canberra: Ronin Films.

Johnston, J. and Baumann, S. (2015). *Foodies: democracy and distinction in the gourmet foodscape* (2nd edn). London: Routledge.

Kaplan, D. H. and Recoquillon, C. (2014). Ethnic place identity within a Parisian neighborhood. *Geographical Review*, 104(1), 33–51.

Kaplan, D. H. and Recoquillon, C. (2015). Multiethnic economic activity along three immigrant corridors in Paris. *The Professional Geographer*, 1–10.

Karvelas, P. and Rushton, G. (2014). Redfern apartment marketer spruiks 'Aboriginal exit'. *The Australian*. Retrieved from www.theaustralian.com.au/business/property/ redfern-apartment-marketer-spruiks-aboriginal-exit/news-story/18f32de355d35556 7302744947ef300b.

Kasinitz, P., Zukin, S. and Chen, X. (2015). Local shops, global streets. In S. Zukin, P. Kasinitz and X. Chen (Eds.), *Global cities, local streets: everyday diversity from New York to Shanghai* (pp. 195–206). New York: Routledge.

Kerwin, D. (2010). *Aboriginal dreaming paths and trading routes: the colonisation of the Australian economic landscape*. Brighton: Sussex Academic.

Khatib, A. (1958). Attempt at a psychogeographical description of Les Halles. *Internationale situationniste 2*. Retrieved from www.cddc.vt.edu/sionline/si/leshalles. html.

Khomami, N. and Halliday, J. (2015). Shoreditch cereal killer cafe targeted by anti-gentrificaton protesters. *Guardian*, 27 September. Retreived from www. theguardian.com/uk-news/2015/sep/27/shoreditch-cereal-cafe-targeted-by-anti-gentrification-protesters.

Khoo, R. (2012). A postcard from Amsterdam, May 8. Retrieved from www. rachelkhoo.com/travel/a-postcard-from-amsterdam.

Kingston, B. (1994). *Basket, bag and trolley: a history of shopping in Australia*. Melbourne and Oxford: Oxford University Press.

Knox, P. L. (2005). Creating ordinary places: slow cities in a fast world. *Journal of Urban Design*, 10(1), 1–11.

Kofman, E. and Lebas, E. (1996). Lost in transposition – time, space and the city. In H. Lefebvre, *Writings on cities* (pp. 3–60). Oxford: Blackwell Publishers.

Kotányi, A. and Vaneigem, R. (1961). Basic program of the Bureau of Unitary Urbanism. *Internationale Situationniste 6*. Retrieved from www.bopsecrets.org/SI/6. unitaryurb.htm.

Krase, J. (2012). *Seeing cities change: local culture and class*. Farnham: Ashgate.

Kunzru, H. (2005). Market forces. *Guardian*, December 7. Retrieved from www. guardian.co.uk/lifeandstyle/2005/dec/07/foodanddrink.shopping.

Lakoff, G. and Johnson, M. (2003). *Metaphors we live by*. Chicago and London: University of Chicago Press.

Landry, C. (2000) *The creative city*. London: Earthscan.

Langer, P. (1984). Sociology – four images of organized diversity: bazaar, jungle, organism, and machine. In L. Rodwin and R. M. Hollister (Eds.), *Cities of the mind: images and themes of the city in the social sciences* (pp. 97–117). New York: Plenum Press.

Laporte, D. (2000). *History of shit*. Cambridge, MA: MIT Press.

Latour, B. (2010). An attempt at a 'Compositionist Manifesto'. *New Literary History*, 41(3), 471–490.

Lavery, S. H. (2014). San Francisco's healthy corner store movement: getting it right, July 4. Retrieved from http://civileats.com/2014/07/04/san-franciscos-healthy-corner-store-movement-getting-it-right/#more-20373.

Law, L. (2011). The ghosts of White Australia: excavating the past(s) of Rusty's Market in tropical Cairns. *Continuum*, 25(5), 669–681.

Le Corbusier (1925). *Plan voisin, Paris, France*. Retrieved from www.fondationlecorbusier.fr/corbuweb/morpheus.aspx?sysId=13&IrisObjectId=6159&sysLanguage=en-en&itemPos=150&itemCount=215&sysParentId=65&sysParentName=Home.

Lefebvre, H. (1991). *The production of space*. Oxford: Blackwell.

Lefebvre, H. (1996). *Writings on cities*. Trans. E. Kofman and E. Lebas. Oxford: Blackwell.

Lentin, A. (2011). Racism in a post-racial Europe. *Eurozine*, November 24. Retrieved from www.eurozine.com/articles/2011-11-24-lentin-en.html.

Lewis, T. and Potter, E. (2011). *Ethical consumption: a critical introduction*. London: Routledge.

Ley, D. (2003). Artists, aestheticisation and the field of gentrification. *Urban Studies*, 40(12), 2527–2544.

Lindner, C. and Meissner, M. (2015). Slow art in the creative city: Amsterdam, street photography, and urban renewal. *Space and Culture*, 18(1), 4–24.

Little book of Sydney villages (n.d.). Retrieved from www.cityofsydney.nsw.gov.au/explore/places-to-go.

L'office du tourisme et des congrès de Paris (2015). *Tourism in Paris: key figures*.

Love, A. (2010). *Broadway Market* [video file], February 2. Retrieved from www.youtube.com/watch?v=_QPNMDTDjQ4.

Low, S. and Smith, N. (Eds.) (2005). *The politics of public space*. New York: Routledge.

Lowy, F. (2006). Regional shopping centres. *Australian Property Journal*, 39(4), 278–283.

Lydon, M. and Garcia, A. (2015). *Tactical urbanism: short-term action for long-term change*. Washington, DC: Island Press.

Lynch, K. (1960). *The image of the city*. Cambridge, MA: MIT Press.

Lynn, G. (2012). Cane rat meat 'sold to public' in Ridley Road Market. *BBC News*. Retrieved from www.bbc.com/news/uk-england-london-19622903.

Lyon, D. and Back, L. (2012). Fishmongers in a global economy: craft and social relations on a London market. *Sociological Research Online*, 17(2), 23.

MacKenzie, D. A., Muniesa, F. and Siu, L. (2007). *Do economists make markets? On the performativity of economics*. Princeton, NJ: Princeton University Press.

MacNaughton, W. (2014). *Meanwhile in San Francisco: the city in its own words*. San Francisco: Chronicle Books.

Maisel, R. (1974). The flea market as an action scene. *Journal of Contemporary Ethnography*, 2(4), 488–505.

Marinelli, M. (2015). Hong Kong street markets as living heritage. *International Institute of Asian Studies Newsletter*, 70(44). Retrieved from www.iias.nl/sites/default/files/IIAS_NL70_4445454647.pdf.

Market Hall (n.d.) Retrieved from www.mvrdv.nl/projects/markethall.

Markowitz, L. (2010). Expanding access and alternatives: building farmers' markets in low-income communities, food and foodways: explorations in the history and culture of human nourishment. *Food and Foodways*, 18(1–2), 66–80.

Massey, D. B. (1994). *Space, place and gender*. Cambridge: Polity.

Massey, D. B. (2005). *For space*. London: Sage.

Massey, D. B., Allen, J. and Pile, S. (1999). *City worlds*. London: Routledge in association with the Open University.

Mayes, C. (2014). An Agrarian imaginary in urban life: cultivating virtues and vices through a conflicted history. *Journal of Agricultural and Environmental Ethics*, 27(2), 265–286.

Mayhew, H. (1861). *London labour and the London poor: Vols. 1–3*. London: Griffin, Bohn, & Company.

McCrohan, D. and Eimer, D. (2015). *Beijing*. Footscray: Lonely Planet.

McDowell, L. (1997). *Undoing place? A geographical reader*. London: Arnold.

McEnearney, M. (n.d.). Carriageworks farmers markets. Retrieved from http://carriageworks.com.au/?page=Event&event=CARRIAGEWORKS-FARMERS-MARKET.

McLennan, G. (2004). Travelling with vehicular ideas: the case of the third way. *Economy and Society*, 33(4), 484–499.

McRobbie, A. (2004). Making a living in London's small-scale creative sector. In D. Power and A. J. Scott (Eds.), *Cultural industries and the production of culture* (pp. 130–143). London: Routledge.

Mead, C. C. (2012). *Making modern Paris: Victor Baltard's Central Markets and the urban practice of architecture*. University Park, PA: Pennsylvania State University Press.

Meek, J. (2011). In Broadway Market [blog post]. *London Review of Books*, August 9. Retrieved from www.lrb.co.uk/blog/2011/08/09/james-meek/in-broadway-market.

Mehta, V. (2013). *The street: a quintessential social public space*. London and New York: Routledge.

Merrifield, A. (2013). *The politics of the encounter: urban theory and protest under planetary urbanization*. Athens, GA: University of Georgia Press.

Merrifield, A. (2014). *The new urban question*. London: Pluto Press.

Metro US (2015). *Christopher Columbus of Brooklyn* [video file], September 25. Retrieved from www.youtube.com/watch?t=1&v=dMij_0q258c.

Miles, S. (2010). *Spaces for consumption: pleasure and placelessness in the post-industrial city*. Los Angeles and London: Sage.

Moreton-Robinson, A. (2003). I still call Australia home: indigenous belonging and place in a white postcolonizing society. In A. M. Fortier, S. Ahmed, C. Castaneda and M. Sheller (Eds.), *Uprootings/regroundings: questions of home and migration* (pp. 23–40). Oxford: Berg.

Morgan, G. (2012). Urban renewal and the creative underclass: Aboriginal youth subcultures in Sydney's Redfern-Waterloo. *Journal of Urban Affairs*, 34(2), 207–222.

Mörtenböck, P. and Mooshammer, H. (2008). Spaces of encounter: informal markets in Europe. *Architectural Research Quarterly*, 12, 347–357.

Mould, O. (2015). *Urban subversion and the creative city*. London: Routledge.

Mumford, L. (1991 [1961]). *The city in history: its origins, its transformations, and its prospects*. London: Penguin Books.

Mundine, M. (2008). On the streets of Redfern there's a new day rising. *Time Out Sydney*, 21 May. Retrieved from www.redwatch.org.au/media/080521tob.

Murphy, K. (producer) (2015). *Life inside the markets* [television series]. Sydney: Seven Network.

National Count of Farmers Market Directory Listing Graph: 1994–2014 (n.d.). Retrieved from www.ams.usda.gov/AMSv1.0/ams.fetchTemplateData.do?template =TemplateS&leftNav=WholesaleandFarmersMarkets&page=WFMFarmersMark etGrowth&description=Farmers+Market+Growth.

Neuwirth, R. (2012). *Stealth of nations: the global rise of the informal economy*. New York: Anchor.

Norberg-Schulz, C. (1980). *Genius loci: towards a phenomenology of architecture*. London: Academy Editions.

Oldenburg, R. (1999). *The great good place: cafés, coffee shops, bookstores, bars, hair salons, and other hangouts at the heart of a community*. New York: Marlowe.

Oliven, R. G. and Pinheiro-Machado, R. (2012). From 'country of the future' to emergent country: popular consumption in Brazil. In J. Sinclair and A. C. Pertierra (Eds.), *Consumer culture in Latin America* (pp. 53–65). New York: Palgrave Macmillan.

Otterloo, A. H. (2009). Eating out 'ethnic' in Amsterdam from 1920s to the present. In L. Nell and J. Rath (Eds.), *Ethnic Amsterdam: immigrants and urban change in the twentieth century* (pp. 41–60). Amsterdam: Amsterdam University Press.

Our history (n.d.). Retrieved from https://web.archive.org/web/20120326065602/http:// www.silkstreet.cc/templet/en/ShowClassPage.jsp?id=ab.

Our introduction (n.d.). Retrieved from https://web.archive.org/web/20080518164517/ http://www.silkstreet.cc/templet/en/ShowClassPage.jsp?id=oi.

Packer, G. (2013). Change the world. *New Yorker*, May 27. Retrieved from www. newyorker.com/magazine/2013/05/27.

Palmer, A. and Clark, H. (Eds.) (2005). *Old clothes, new looks: second hand fashion*. Oxford: Berg.

Parham, S. (2012). *Market place: food quarters, design and urban renewal in London*. Newcastle: Cambridge Scholars Publishing.

Parham, S. (2015). *Food and urbanism: the convivial city and a sustainable future*. London and New York: Bloomsbury.

Paul Keating's Redfern speech [video file]. (2012). Retrieved from http://antar.org.au/ reports/paul-keatings-redfern-speech.

Peck, J. (2005). Struggling with the creative class. *International Journal of Urban and Regional Research*, 29(4), 740–770.

Peck, J. (2012). Recreative city: Amsterdam, vehicular ideas and the adaptive spaces of creativity policy. *International Journal of Urban and Regional Research*, 36(3), 462–485.

Peel Street/Graham Street Redevelopment Scheme (n.d.). Retrieved from www.ura.org.hk/ en/projects/redevelopment/central/peel-street-graham-street-development-scheme. aspx.

Perry, G. (2014). *Metropolitan meat market* [video file], March 23. Retrieved from www.youtube.com/watch?t=30&v=gb2UxaodZj0.

Petrini, C. (2003). *Slow food: the case for taste*. New York: Columbia University Press.

Petrini, C. (2007). *Slow food nation: why our food should be good, clean, and fair*. New York: Rizzoli.

Pink, S. and Mackley, K. L. (2014). Moving, making and atmosphere: routines of home as sites for mundane improvisation. *Mobilities*, 1–17.

Pirenne, H. (2014 [1925]). *Medieval cities: their origins and the revival of trade.* Princeton, NJ: Princeton University Press.

Plattner, S. (1982). Economic decision making in a public marketplace. *American Ethnologist*, 9(2), 399–420.

Plattner, S. (1985). Markets and marketing. *Current Anthropology*, 26(3), 387–389.

Plumwood, V. (2008). Shadow places and the politics of dwelling. *Australian Humanities Review*, 44. Retrieved from www.australianhumanitiesreview.org/archive/Issue-March-2008/plumwood.html.

Pratt, A. C. (2011). The cultural contradictions of the creative city. *City, Culture and Society*, 2(3), 123–130.

Project for Public Spaces (2003). *Public markets and community-based food systems: making them work in lower-income neighbourhoods.* New York: Authors.

Public Markets (n.d.) Retrieved from www.pps.org/markets.

Rabbiosi, C. (2015). Renewing a historical legacy: tourism, leisure shopping and urban branding in Paris. *Cities*, 42(B0), 195–203.

Randwick City Council (2014, May 20). *Light rail to Randwick – Kingsford car park* [video file]. Retrieved from www.youtube.com/watch?v=LfNr1Z92r_w.

Rasmussen, S. E. (1982 [1934]). *London: the unique city* (Rev. edn). Cambridge, MA: MIT Press.

Ratcliffe, B. (1992). Perceptions and realities of the urban margin: the rag pickers of Paris in the first half of the nineteenth century. *Canadian Journal of History*, 27(2), 197–233.

Redfern Night Markets (n.d.) Retrieved from http://redfernnightmarkets.com.au.

Redfern-Waterloo Authority (n.d.) *RWA Submission: Sustainable Sydney 2030 Final Consultation Draft.* Retrieved from www.redfernwaterloo.com.au/other/rwa_submission.pdf.

Reimerink, L. (2015). Amsterdam to tourists: get off the beaten path. *Citiscope*, March 26. Retrieved from http://citiscope.org/story/2015/amsterdam-tourists-get-beaten-path.

Rhys-Taylor, A. (2013). The essences of multiculture: a sensory exploration of an inner-city street market. *Identities*, 20(4), 393–406.

Rhys-Taylor, A. (2014). Intersemiotic fruit: mangoes, multiculture and the city. In H. Jones and E. Jackson (Eds.), *Stories of cosmopolitan belonging: emotion and location* (pp. 44–56). Abingdon: Routledge.

Ridley Road rat scandal [blog post] (2012). Retrieved from http://broadwaymarket.co.uk/index.php.

Roy, A. (2011). Slumdog cities: rethinking subaltern urbanism. *International Journal of Urban and Regional Research*, 35(2), 223–238.

Rushton, G. (2014). Redfern apartment marketer spruiks 'Aboriginal exit'. *Australian*, 10 December. Retrieved from www.theaustralian.com.au/business/property/redfern-apartment-marketer-spruiks-aboriginal-exit/story-fn9656lz-1227150438149.

Sánchez, F. and Broudehoux, A.-M. (2013). Mega-events and urban regeneration in Rio de Janeiro: planning in a state of emergency. *International Journal of Urban Sustainable Development*, 5(2), 132–153.

Scanlan, J. (2005). *On garbage.* London: Reaktion.

Seale, K. (2009). *Textual refuse: Iain Sinclair's politics and poetics of refusal.* Doctoral thesis, University of Sydney.

Seale, K. (2012). MasterChef's amateur makeovers. *Media International Australia incorporating Culture and Policy*, 143, 28–35.

Seale, K. (2014). On the beach: informal street vendors and place in Copacabana and Ipanema, Rio de Janeiro. In C. Evers and K. Seale (Eds.), *Informal urban street markets: international perspectives* (pp. 83–94). New York: Routledge.

Seale, K. and Evers, C. (2014). Informal urban street markets: international perspectives. In C. Evers and K. Seale (Eds.), *Informal urban street markets: international perspectives* (pp. 1–16). New York: Routledge.

Secchi, B. and Viganò, P. (2009). *Antwerp, territory of a new modernity*. Amsterdam: SUN.

Sennett, R. (1993 [1974]). *The fall of public man*. London: Faber.

Shaw, W. S. (2007). *Cities of whiteness*. Oxford: Blackwell.

Shaw, W. S. (2013). Redfern as the heart(h): living (black) in inner Sydney. *Geographical Research*, 51(3), 257–268.

Shields, R. (1991). *Places on the margin: alternative geographies of modernity*. London: Routledge.

Shortell, T. and Brown, E. (Eds.) (2014). *Walking in the European city: quotidian mobility and urban ethnography*. Farnham: Ashgate.

Simms, A. (2014). Urban corporate governance and the shaping of medieval towns. In M. Conzen and P. Larkham (Eds.), *Shapers of urban form: explorations in morphological agency* (Vol. 63–80). New York: Routledge.

Sinclair, I. (1995). *White Chappell, scarlet tracings*. London: Vintage.

Sinclair, I. (1997). *Lights out for the territory*. London: Granta.

Sinclair, I. (2006). Lost treasure. *Guardian*, March 18. Retrieved from www.theguardian.com/society/2006/mar/18/communities.weekendmagazine.

Sinclair, I. (2009). *Hackney, that rose-red empire*. London: Penguin.

Siradeau, S. (2008). *Vintage French interiors: inspiration from the antique shops and flea markets of France*. Paris: Flammarion.

Slater, D. and Tonkiss, F. (2001). *Market society: markets and modern social theory*. Cambridge: Polity.

Slater, T. R. (2014). Urban corporate governance and the shaping of medieval towns. In M. Conzen and P. Larkham (Eds.), *Shapers of urban form: explorations in morphological agency* (pp. 46–62). New York: Routledge.

Smith, N. (1996). *The new urban frontier: gentrification and the revanchist city*. London: Routledge.

Smith, N. (2002). New globalism, new urbanism: gentrification as global urban strategy. *Antipode*, 34(3), 427–450.

Solnit, R. and Schwartzenberg, S. (2002). *Hollow city: the siege of San Francisco and the crisis of American urbanism*. London and New York: Verso.

Solnit, R., Pease, B. and Seigel, S. (2010). *Infinite city: a San Francisco atlas*. Berkeley, CA: University of California Press.

Sørensen, S. B. (producer) and Dalsgaard, A. (director) (2012). *The human scale*. Copenhagen: Final Cut for Real.

Stallybrass, P. and White, A. (1986). *The politics and poetics of transgression*. London: Methuen.

SubaruAustralia (2014). *River Cottage Australia, Eveleigh Markets Official Subaru Australia* [video file]. Retrieved from www.youtube.com/watch?v=tMepysRs4mQ.

Sully, A. and Westbury, M. (writers), and Venables, A. (producer) (2015). *Bespoke* [television series]. Viking Films, Australian Broadcasting Corporation & Screen Australia.

Sunday Shopping (n.d.) Retrieved from www.visitantwerpen.be/detail/sunday-shopping.

Tangires, H. (2008). *Public markets*. New York: W.W. Norton in association with Library of Congress.

Tantao News (2010). *Beijing Silk Market manager arrested* [video file]. Retrieved from www.youtube.com/watch?v=74KIWmL1L7U.

Taylor, C. (2004). *Modern social imaginaries*. Durham, NC: Duke University Press.

Taylor, C. (2011). *Londoners : the days and nights of London now – as told by those who love it, hate it, live it, left it and long for it*. London: Granta.

Ten Hoor, M. (2007). Architecture and biopolitics at Les Halles. *French Politics, Culture & Society*, 25(2), 73–92.

Tonkiss, F. (2005). *Space, the city and social theory: social relations and urban forms*. Cambridge: Polity.

Tonkiss, F. (2013). *Cities by design: the social life of urban form*. Cambridge: Polity.

Top 100 City Destinations Ranking (2015). Retrieved from http://blog.euromonitor.com/2015/01/top-100-city-destinations-ranking.html.

Trauger, A. (Ed.) (2015). *Food sovereignty in international context: discourse, politics and practice of place*. Abingdon: Routledge.

Trotter, D. (2000). *Cooking with mud: the idea of mess in nineteenth-century art and fiction*. Oxford: Oxford University Press.

United Nations Human Settlements Programme. (2010). *Solid waste management in the world's cities: water and sanitation in the world's cities 2010*. London and Washington, DC: UN Habitat/Earthscan.

Urry, J. (2002). *The tourist gaze*. London: Sage.

Valentine, G. (2008). Living with difference: reflections on geographies of encounter. *Progress in Human Geography*, 32(3), 323–337.

Veeck, A. (2000). The revitalization of the marketplace: the food markets of Nanjing. In D. Davis (Ed.), *The consumer revolution in urban China* (pp. 107–123). Berkeley, CA: University of California Press.

Veeck, G., Veeck, A. and Zhao, S. (2015). Perceptions of food safety by urban consumers in Nanjing, China. *The Professional Geographer*, 67(3), 490–501.

Venkatesh, S. A. (2006). *Off the books: the underground economy of the urban poor*. Cambridge, MA: Harvard University Press.

Venkatesh, S. A. (2008). *Gang leader for a day: a rogue sociologist crosses the line*. London: Allen Lane.

Visconti, L. M., Minowa, Y. and Maclaran, P. (2014). Public markets: an ecological perspective on sustainability as a megatrend. *Journal of Macromarketing*, 34(3), 349–368.

Visit Antwerpen (n.d). Retrieved from: http://web.archive.org/web/20150205144643/http://www.visitantwerpen.be/bze.net?id=1470.

Wacquant, L. (2008). *Urban outcasts: a comparative sociology of advanced marginality*. Cambridge: Polity.

Wainwright, O. (2014). Rotterdam's Markthal: superdutch goes supersized in psychedelic marketplace. Retrieved from www.theguardian.com/artanddesign/architecture-design-blog/2014/oct/02/-sp-rotterdam-markthal-superdutch-market-mvrdv.

Wakeman, R. (2007). Fascinating Les Halles. *French Politics, Culture and Society*, 25(2), 46–72.

Watson, S. (2006). *City publics: the (dis)enchantments of urban encounters*. Abingdon: Routledge.

Watson, S. (2009). The magic of the marketplace: sociality in a neglected public space. *Urban Studies*, 46(8), 1577–1591.

Watson, S. and Studdert, D. (2006). *Markets as sites of social interaction: spaces of diversity*. London: Joseph Rowntree Foundation and Policy Press.

Watson, S. and Wells, K. (2005). Spaces of nostalgia: the hollowing out of a London market. *Journal of Social and Cultural Geography*, 6 (6), 17–29.

Welcome (n.d.). Retrieved from http://web.archive.org/web/20110823214741/http://www.broadwaymarket.co.uk.

Wessendorf, S. (2013). Commonplace diversity and the 'ethos of mixing': perceptions of difference in a London neighbourhood. *Identities*, 20(4), 407–422.

Wessendorf, S. (2014). *Commonplace diversity: social relations in a super-diverse city*. New York and London: Palgrave Macmillan.

Whyte, W. H. (director) (1988). *Social life of small urban spaces* [motion picture]. USA: Direct Cinema Limited.

Wirth, L. (1938). Urbanism as a way of life. *The American Journal of Sociology*, 44(1), 1–24.

Wolford, W. (2004). This land is ours now: spatial imaginaries and the struggle for land in Brazil. *Annals of the Association of American Geographers*, 94(2), 409–424.

Wong, A. (2015). Central Market – a modern-day market for Hong Kong. *Hong Kong Free Press*, July 3. Retrieved from www.hongkongfp.com/2015/07/03/central-market-a-modern-day-market-for-hong-kong.

Wood, E. M. (2002). *The origin of capitalism*. London: Verso.

Wright, P. (2009). *A journey through ruins: the last days of London*. Oxford and New York: Oxford University Press.

Young, C., Karpyn, A., Uy, N., Wich, K. and Glyn, J. (2011). Farmers' markets in low income communities: impact of community environment, food programs and public policy. *Community Development*, 42(2), 208–220.

Young, I. M. (1990). *Justice and the politics of difference*. Princeton, NJ: Princeton University Press.

Zepeda, L. and Reznickova, A. (2013). *Measuring effects of mobile markets on healthy food choices*. University of Wisconsin. Retrieved from http://dx.doi.org/10.9752/MS142.11-2013.

Zola, E. (2007 [1883]). *Belly of Paris*. Oxford: Oxford University Press.

Zucker, P. (1970). *Town and square*. Cambridge, MA: MIT Press.

Zukin, S. (1991). *Landscapes of power: from Detroit to Disney World*. Berkeley, CA: University of California Press.

Zukin, S. (1995). *The cultures of cities*. Oxford: Blackwell.

Zukin, S. (2008). Consuming authenticity. *Cultural Studies*, 22(5), 724–748.

Zukin, S. (2012). The social production of urban cultural heritage: Identity and ecosystem on an Amsterdam shopping street. *City, Culture and Society*, 3(4), 281–291.

Zukin, S., Kasinitz, P. and Chen, X. (Eds.) (2015). *Global cities, local streets: everyday diversity from New York to Shanghai*. New York: Routledge.

Index

Printed and bound by CPI Group (UK) Ltd, Croydon, CR0 4YY

22/10/2024

01777628-0016